VIS Radom

A Study and Photographic Album of Poland's Finest Pistol

William J. York

Published and printed in the United States of America

Commercial photography and cover photograph by Anthony Vanderlinden - Wet Dog Publications
Graphic design by Abbie Enk

Wet Dog Publications
5603-B West Friendly Avenue, Suite 166
Greensboro, North Carolina 27410
USA

ISBN: 978-0-9981397-2-2

Acknowledgments

Leonardo M. Antaris, MD for sharing information on the Ruby .45 and contributing photographs.

Julian Daszkowski of Warsaw, Poland obtained a history of the Radom factory for the author.

Guido Davids for providing translations of the German documents.

William Drake for sharing his knowledge on the subject with the publisher.

Francis Kennedy for information and access to his collection of VIS pistols.

Peter G. Kokalis, for his friendship and encouragement.

M. Peter Kuck, past President of Ye Connecticut Gun Guild, for his friendship and the loan of a German "tobacco can" cleaning kit.

Tom Nelson, who shared his years worth of knowledge with the author.

Ron Pinkenburg for access to his collection of VIS pistols.

David Rachwal for access to a BNZ pistol.

Rock Island Auction Company for contributing photographs of the presentation pistol.

Edward Tytor, President of Ye Connecticut Gun Guild and collector of Polish firearms, for access to his collection of VIS pistols, and for sharing much knowledge with the author.

Waclaw Ustupski, for opening his vast collection and database to the author, and for contributing factory documents and photographs for this book. His part was that of encourager, prime mover, and collaborator, without whose help this book would hardly have been possible to create.

Anthony Vanderlinden, who was both the publisher and a collaborator in this work.

Abbie Enk for select graphic design elements.

Select acknowledgements from contributor Waclaw Ustupski:

Zbigniew Gwozdz, Poland based firearms author and historian, for photographs and materials from Radom.

Alicja Kopycka, who photographed the memorial at the Radom factory.

Marta Nemoda, who helped gain access to the .22 VIS in the Budapest Museum.

Steve Vandtke, who allowed his unfinished, Kriegsmarine, and Yugoslav VIS pistols to be photographed.

Emil Walega, veteran anti-German partisan and Poland based arms collector.

Anna Ustupska Chyc, wife and collaborator.

Content

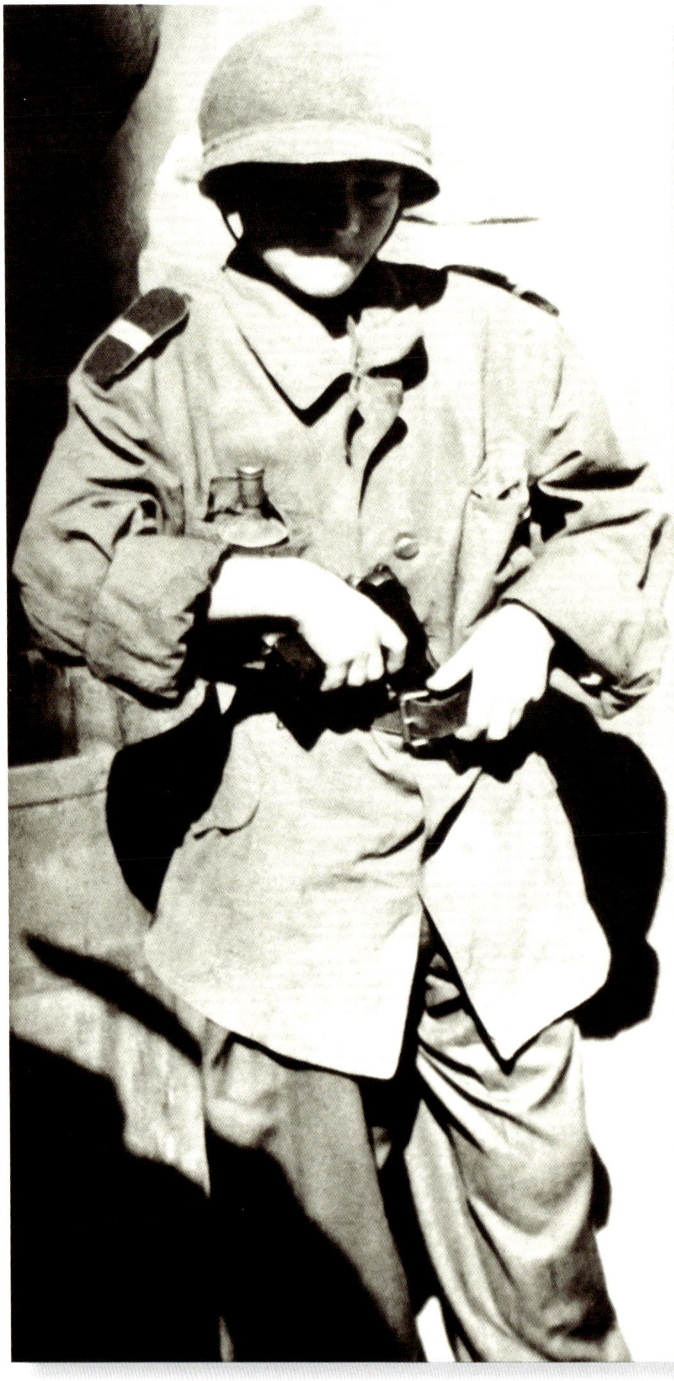

(Fig.) A young boy holding a VIS, photograph most likely taken during the Warsaw Uprising. (Photograph courtesy of Waclaw Ustupski)

Introduction

The Polish VIS pistol, popularly known as the Radom (after the place of manufacture stamped as a part of its slide legend) has been popular among U.S. gun collectors ever since the first souvenir examples made their appearance in the aftermath of the Second World War. It was first mentioned in the United States in a November 1944, article by W.H.B. Smith in The American Rifleman. Since that time, collectors have been increasingly interested in both the pre-war and wartime variations of this weapon, aided by an increasing body of knowledge of its variations and history. The VIS pistol is, in fact, a product of the pre-war "golden age" of automatic pistol design. It is reliable and well known for its rugged strength, and may still be often seen on the shooting range.

More than this, however, it is the wealth of variety seen in the VIS, from beautifully finished pre-war examples to extremely crude but still functional late war specimens, that has attracted collectors, from the beginner to the advanced, to the study of this pistol.

The author of this book was likewise inspired by this wealth of variety, and set out to outline the history of the VIS, to put forth some of the systems used in cataloging its variations, and to provide as extensive as possible a photographic catalog of the VIS. It is hoped that he has, in some way, succeeded.

Special thanks are due to my publisher and photographer, Anthony Vanderlinden, to Waclaw Ustupski-Chyc of Illinois, an eminent collector, to Edward Tytor, friend and avid collector, as well as President of Ye Connecticut Gun Guild, and to my circle of friends who encouraged this work. All errors and omissions are, alas and of course, my own.

William J. York

Chapter I
Poland's Pistol - A Historical Overview

A Brief Background on its Country of Origin

Poland, united in commonwealth with Lithuania, was at one time the largest country in Europe. Her armies once occupied Moscow, and, in 1683, Polish forces lifted the Turkish siege of Vienna, thereby ending the era of Turkish expansion in Europe.

The old Polish commonwealth was, however, severely hampered by its largely ineffective form of central government. An elected monarch ruled the country at pretty much the sufferance of the Sejm, or parliament (today's Polish parliament is also called a Sejm). All members of the nobility could claim a seat in the old commonwealth's Sejm, and since Poland had the greatest percentage of nobles in all of Europe, some 10 percent of its population (many penniless), parliamentary sessions could be long, and unruly. Unless, of course, a veto was cast.

The peculiar system of veto in the old Polish Sejm was called the LIBERUM VETO, and was the worst aspect of the system. Under it, any member of parliament was able to put a halt to legislation by standing up and saying " I will not permit it". Once the veto was cast, the Sejm was required to adjourn, and its members would return home. This system was constantly subject to bribery and intimidation, with the result that Poland's central government was powerless to rule the country. The result of all this was a "noble anarchy" of the wealthiest nobles, known as magnates.

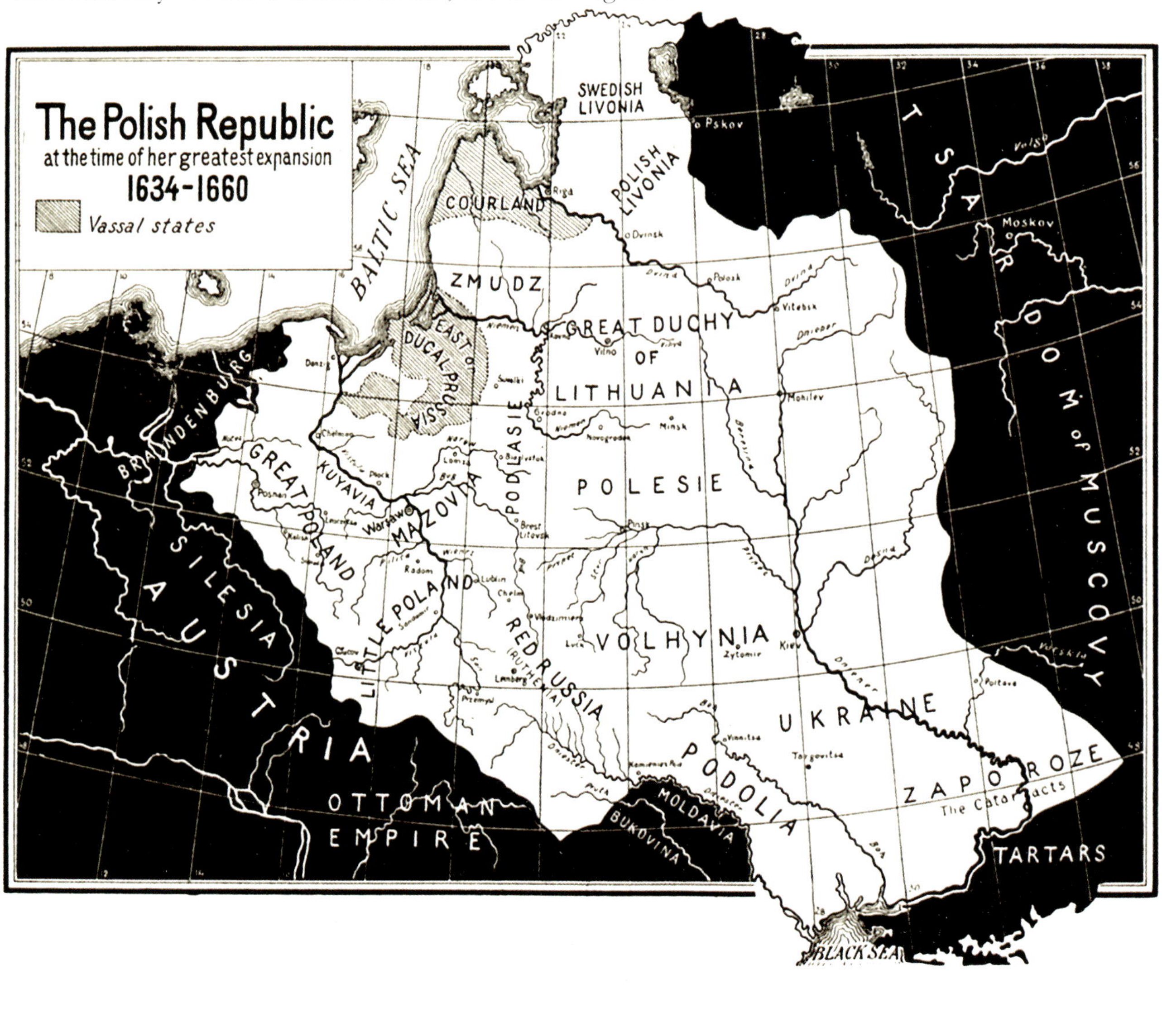

The elected nature of Poland's kings posed another problem, as it was constantly subjected to the intrigues, influences, and bribery of contemporary world powers, and especially by Poland's neighbors. Eventually and inevitably, Poland's political weakness led to her dismemberment by her neighbors Austria, Prussia, and Russia, which dismemberment took place in three partitions of 1772, 1793, and 1795. Poland's adoption of a liberal constitution in 1791 did nothing to aid her survival.

The proud Poles did not simply accept partition and foreign rule. The passing years were to bring a history of uprisings and failed governments on the territory of the former Poland. The Prussians tried to make their Polish subjects into Prussians, the Russians into Russians. Only the Austro-Hungarians left the Poles more or less alone. And so it went, up to the great upheaval of the Twentieth Century, the First World War.

Polish forces fought on both sides of World War I, their leaders hoping for the long-awaited national rebirth in the war's aftermath. As a result of Russia's retreat out of Polish territory, Germany was able, during the pendency of hostilities, to organize a provisional government of a new Kingdom of Poland in Warsaw. Upon the subsequent collapse of the Central Powers, Poland was reborn as a Republic on November 11, 1918.

The politics and leadership of post-1918 Poland, as well as a study of the personalities involved, constitute a fascinating field of study, but this would take us too far afield. Rather, let us move to a consideration of the strategic military situation faced by the new Poland.

Military Challenges: 1918-1922

In 1918, the newly reborn Poland started out with three different regional and local systems of government, three separate economic systems, and uncertain national borders. Her armed forces were formed from units and veterans of most of the World War's combatant states, and her armaments consisted of whatever she had inherited from these combatants. The result of this situation would be a logistical nightmare in the best of times. For the new Poland, these were anything but the best of times.

Poland's western territories consisted in large part of lands previously under Prussian control. One important part of this territory was Upper Silesia, which had an ethnic Polish majority. After a series of bloody uprisings in support of Polish rule, a plebiscite was finally held in this region, the result being a majority vote to remain a part of Germany!

Since Germany was, in the aftermath of the war, too weak to do anything about it, Poland took over in Upper Silesia notwithstanding the results of the plebiscite, causing bitter feelings with the German state. Of course, Germany was also unhappy with the Treaty of Versailles' creation of a "corridor", which gave Poland access to the Baltic Sea, but which also had the result of cutting off East Prussia from the rest of Germany. These matters simmered until 1939, when a newly strong Germany began World War II by invading Poland.

In the South, both Poland and the equally new nation of Czechoslovakia claimed the small mining area known to the Poles as Zaolzie or Cieszyn. The population of this area was by majority Polish, but Czech diplomats managed to avoid a plebiscite, and Czechoslovakia took the territory. A flipside, perhaps, of the Upper Silesian situation, but one which caused bitter feelings in Poland. Poland was to join with Germany in 1938 to (temporarily) reclaim this territory.

To Poland's northeast lay the newly constituted Lithuania. Now, Poland and Lithuania had long been united in a past Commonwealth, and the Lithuanian city of Vilnius (Wilno in Polish) not only had a majority Polish population, but was the birthplace of Josef Pilsudski, Poland's strongman and military leader. In 1920, in the aftermath of Poland's victory over Soviet forces (see the following paragraph), Polish forces occupied the Wilno region, which Poland subsequently annexed. Needless to say, this caused bitter and strained relations with Lithuania which almost led to war.

To Poland's East, was a long border with lands populated by Ukrainians and Byelorussians, which lands had long been a part of the Russian Empire, and which were now claimed by Russia's new Bolshevik government. There were Polish minorities living in these areas, and Poland became, in 1920, involved in a war with the Soviet Union. In this war, attacking Soviet forces were to reach the eastern suburbs of Warsaw when, in the "miracle of the Vistula", Polish forces under Pilsudski broke through the Bolsheviks' weak center, and cut off and decimated the invader's overextended armies from the rear. The resulting peace treaty awarded Poland a large eastern territory which, while historically Polish, now contained a majority non-Polish population. The outcome of this war also created bad relations with the USSR, and was to give Stalin a pretext to plot with Hitler in 1939, leading to Poland's loss of the eastern third of her territory.

Clearly, the new Poland was faced with the fact that her two largest neighbors were unfriendly ones. She needed a well-equipped military.

Poland's Industrial and Small Arms Situation

Due to the circumstances of her rebirth, Poland's forces were initially equipped with a confusing array of small arms, both modern and obsolete. The accompanying chart illustrates this situation at two intervals.

Pistols in the Polish Army: 1918-1936

I. List as of Aug. 29, 1922: Total number reported: 17,000 (Number Authorized: 70,000)

Principal weapons reported (no break down given as to quantities):

Mauser M1896
Mauser M1910
Frommer M1911
Luger M1908
Roth-Steyr M1907
Steyr Hahn M1911
Ruby (Cebra) M1916

The most popular officer self-purchased handguns were:
(Officers were free to purchase any pistol in 7.65mm Browning (.32 ACP) or larger caliber.)

Browning M1910
Browning M1922
Mauser M1910
Beretta M1923

The army stocked ammunition in calibers 7.65mm and 9mm Browning Short, 8 and 9mm Steyr, 9mm Parabellum and 7.63mm Mauser. Polish border guard units were armed with 8mm French M1892 revolvers as well, and police were often armed with 7.62 mm gas-seal Nagant revolvers.

II. Additionally, as a matter of necessity, the Polish Armed Forces Ministry published in 1931 a manual of pistols and revolvers which covered:

Revolvers:

Gasser M1898
French "Lebel" M1892 ("two types")
Smith & Wesson M1905
OEE Orbea (8mm French)
Pieper M1889 (8mm French)
Nagant M1895 (7.62mm)

Pistols:

Mauser M1896
Mauser M1910
Frommer (7.65&9mm)
Steyr M1907
Steyr M1912
Parabellum (Luger) M1908 (two types)
Ruby (Cebra) M1916
Dreyse M1907
Pieper M1909
Ortgies
FN Browning M.1900, 1906, 1910, 1922
Bayard M1908

(Fig. 11-1) The above figures are from BRON STRZELECKA I SPRZET ARTYLERYSKI FORMACJI POLSKICH I WOJSKA POLSKIEGO W LATACH 1914-1939, by Andrzej Konstankiewicz, Lublin, Uniwersitet Marii Curie-Sklodowski, 2003; and BRON STRZELECKA WOJSK POLSKICH W LATACH 1717-1945, by Marian Maciejewski, Szczecin, Wydawnictwo Glob, 1991.

Poland's finances did not permit either rapid rearmament or the state-funded development of unique weapons systems, but some resources were nonetheless available to her. The rifle-making machinery from the former Imperial German armory in Danzig was ordered transferred to Warsaw by the administrators of the treaty of Versailles, and this machinery went to the State Rifle Factory (Panstwowa Fabryka Karabinow-P.F.K.) in Warsaw. Later, this equipment was transferred to the town of Radom, to create the arms works there. The government plans for the Radom factory date from 1922, construction started in 1923 and finished in 1927. These works were first called P.W.B. (Panstwowa Wytwornia Broni - State Weapons Factory), and later became F.B. (Fabryka Broni - Weapons Factory) by government decree on April 27, 1927. In addition to its main arsenal in Radom, Poland had inherited a number of repair depots from the World War combatants. These were used for arms maintenance, and also for overhaul and conversion work, an example of the latter being the conversion and re-barreling of Russian Mosin-Nagant rifles.

In the aftermath of Poland's 1920 war against Bolshevik Russia, the Polish military decided to standardize the German 7.92x57mm Mauser round as its rifle caliber. Rifles and carbines could be easily produced in this caliber by means of the ex-Danzig machinery, and a fair number of captured and inherited Mosin-Nagant rifles were converted to this caliber by Polish repair depots as well. Other, incompatible rifles in Poland's inventory were gradually phased out to the reserves or out of use entirely, or were used as trading stock with other countries for more desirable arms.

The pistol situation in the Polish armed forces immediately after independence was, if anything, even more of a jumble than that of rifles. The accompanying chart (fig. 11-1) attests to this.

Rearming with a standardized handgun was of secondary importance to Poland's military planners, but eventually steps were taken to this end. Besides a trade with Yugoslavia for a quantity of Roth-Steyr M1907 Austrian cavalry pistols, Poland acquired a set of Belgian tooling for the manufacture of the Russian model gas-seal Nagant revolver. The latter was put into production at the Radom factory as the model NG30 revolver.

Finally, at the end of the 1920s, the Polish military was ready to move toward the adoption of a new automatic pistol, as an item of general issue in her armed forces. The original proposal on the table was to adopt an imported model, a version of the rotating-barrel Czech VZ24 pistol in caliber 9mm Browning Short (.380 acp). This is where the story of the VIS-Radom pistol begins.

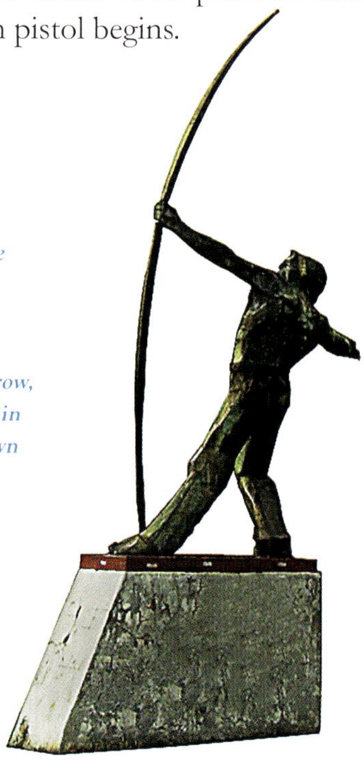

(Fig. 12-1) The Radom arms factory used the symbol of an archer in the prewar and post Communist eras. The word "Arrow" in Polish has the same origins/roots as the words "shoot" or "shot". The origins of these words are the same in Russian and Russian factories like Sestroyetsk, Izhevsk and Tula used similar symbols as logos. Sestroyetsk used an arrow, Izhevsk used a bow and arrow or just an arrow and Tula an arrow within a five pointed star. Seen here is a statue of the archer, located in the town of Radom.

Chapter II
Enter the VIS

The principal protagonist of the pistol which came to be known as the VIS was Piotr Wilniewczyc, who in 1929 was employed at the National Ordnance Company in Warsaw. He was assisted in his efforts by Jan Skrzypinski, an ordnance engineer and director of the State Rifle Factory in Warsaw, who was to join the effort which resulted in the VIS, at an early date. Also involved in the VIS project was Andrzej Dowknott, then technical director of Poland's National Ordnance Company. The story of how the VIS came to be designed was related by Mr. Wilniewczyc in an article entitled VIS I MORS (VIS and MORS) in Volume I of Musealnictwo Wojskowe, Warsaw, 1959. (The English name of this publication is "Army Museology", the MORS is a sub-machine gun also designed by Wilniewczyc, which never reached serial production due to the outbreak of World War II.)

As stated above, the Polish military authorities had in 1929 decided to adopt a version of the Czech Model 1924 pistol as part of a plan to improve Poland's small arms situation. The version considered was known as the VZ 1928. This was a rotating barrel, locked breech design in caliber 9mm Browning Short (.380 ACP), and was a derivation of designs made by the German/Czech Josef Nickl while he was an employee of the Mauser company. The Polish authorities were familiar with this design, as Poland's Ministry of the Treasury had purchased some 1700 of these pistols in 1929 and 1930. The Polish Government had approved entering into a licensing agreement with the Czechs which would allow the manufacture of this pistol in Poland for a royalty payment of some $250,000.00. That is where the VIS story was to begin.

In 1929, when Mr. Dowknott was made aware of the proposal to manufacture the VZ1928 under license, he shared the information with Piotr Wilniewczyc. Underwhelmed by the pistol, feeling that the price was too high, and sure that a better, native design could be put forth on short notice, the two men informed the Polish military authorities that they had been working on a pistol of superior design. They submitted a proposal which had, in actuality, been put together in the space of only two days! This proposal bought enough time for Wilniewczyc to get about the business of designing the pistol that was to become the VIS. In this, he was joined by Jan Skrzypinski, who had been impressed by the preliminary drawings.

The decision of the Polish military authorities to delay things in favor of a native design was doubtless due in part to the secondary role played by the pistol in warfare. The Polish Cavalry, at that time a highly respected part of the armed forces, likely also played a role, in view of the marginal, at best, adequacy of the VZ1928's cartridge. Polish nationalism and an abiding resentment of the Czech state over its takeover of Zaolzie would seem to also have been factors favoring a Polish design.

By the end of 1930, a set of drawings for the preparation of the first VIS prototype were ready. Recently published historical information from the Radom factory has revealed that the final product was prepared by a young engineer named Feliks Modzelewski, who had previously worked on Browning aircraft machinegun designs, and who was to perish in the Second World War. In February,1931, a finished prototype pistol was ready for its initial trials. These drawings and the initial prototype (which bore on its slide the name of the State Rifle Factory in Warsaw), were the basis of an application resulting in the granting of Polish Patent No. 15567 on April 8, 1932. The construction of the prototype was also the occasion for giving the pistol a name. Skrzypinski suggested that the name be "WIS", from the two designers' initials. This was later changed by the Polish military authorities to VIS, which is the Latin word for strength. The name "VIS" can be confusing to those familiar with the Polish language, since the Polish alphabet has no letter "V".

The first VIS prototype proved itself well-made and serviceable, and was to fire 6000 shots with no mechanical breakdowns. Contemporary targets illustrated in Wilniewcyc's article show that his pistol remained remarkably accurate throughout its trials. Interestingly, the Colt M1911 pistol had also fired 6000 shots with no failures in its final pre-adoption trials; a fact much employed in Colt advertising of the day. All was not perfect with the VIS, however, and a number of modifications were made to the pistol.

The VIS as first constructed employed a captive recoil spring on a rigid rod. This was changed to a flexible (and more complicated) design featuring a movable rear section powered by a secondary, internal spring. This was done to prevent the pistol from accidentally firing when dropped on its barrel. A hammer drop lever was later added, to enable the soldier to safely decock the weapon with one hand, a desire of the cavalry. The pistol's sear and disconnector were also modified, in order for the hammer drop to work correctly. The firing pin also required the addition of a flat surface towards its rear, which was cammed by the hammer drop to withdraw the firing pin into the slide and allow the hammer to fall without any danger of firing the pistol.

Other changes to the VIS were a repositioning of its ejector to the center of the slide, changes to the sights, and a reshaping of the rear of the frame and slide for a more comfortable hold. After the first prototypes, the hammer was modified to a rounded "rowel" type , allegedly so a cavalryman could cock the pistol against his boot or saddle. Finally, the pistol's mainspring housing was machined with a slot for the attachment of a shoulder stock.

The Polish military authorities were duly pleased with the VIS. Not only was it reliable, accurate, and safe, but it was a native design and it chambered the 9mm Parabellum cartridge, a round substantially more powerful than the 9mm Browning Short. Rights to the pistol were purchased, and in late 1932 to early 1933 a trial run of some 30, according to some sources, but actually some 134 according to others, of the new pistols was prepared and issued for further testing to several units. These were the 4th, 21st, and 72nd Infantry Regiments, the 6th Uhlan Regiment, the 3^{d} and 10th Regiments of Mounted Riflemen, and the 5th, 21st, and 27th Light Artillery Regiments. These pistols were the first to be produced at Radom, and were so marked; though their markings differed from those on later production pistols. One side of their slides was marked "Panstwowe Wytwornie Uzbrojenia, Fabryka Broni w Radomiu" (State Weapons Factory, Weapons Factory in Radom). The other side was marked "VIS kal.9mm Pat. N° 15567". A quality control mark was placed on the frame, on the trigger guard. These trials ended in success, and series production of the pistol to be known as the "VIS wz.1935" (VIS Model 1935), commenced in 1936. The Polish army's projected pistol requirements at that time were for about 90,000 of the new pistols.

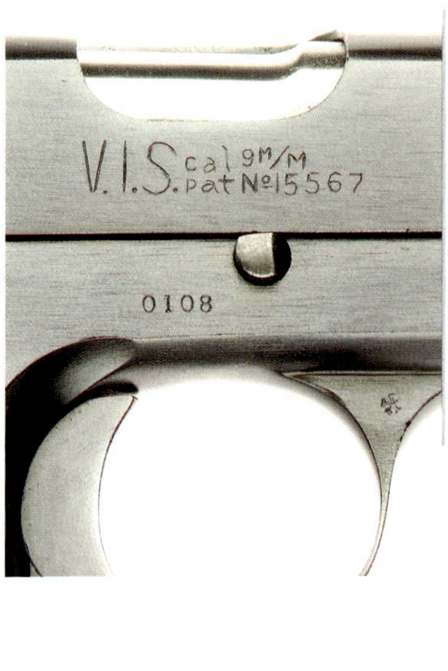

(Fig. 14-1) The left and right side slide markings of prototype VIS, Serial Number 0108. The left side reads " State Weapons Factory, Weapons Factory in Radom" (Photographs by Anthony Vanderlinden, courtesy of Waclaw Ustupski collection)

8 kwietnia 1932 r. 2

F41c 5/00

URZĄD PATENTOWY

RZECZYPOSPOLITEJ POLSKIEJ

OPIS PATENTOWY

Nr 15567. Kl. 72 h 5.

Piotr Wilniewczyc
(Warszawa, Polska)
i Jan Skrzypiński
(Warszawa, Polska).

Pistolet samoczynny.

Zgłoszono 15 stycznia 1931 r.
Udzielono 8 lutego 1932 r.

W celu podniesienia wartości użytkowej pistoletów znanych systemów, niniejszy wynalazek wprowadza szereg nowych urządzeń, wymienionych poniżej.

Omawiany pistolet stanowi broń samoczynną o ryglowanej lufie, przyczem odrzut wykorzystany jest dla odryglowywania lufy i otwierania zamka, wyciągania i wyrzucania łuski oraz napinania kurka.

Po daniu strzału zamek pod działaniem sprężyny wraca do swego normalnego położenia, powodując wprowadzenie z magazynka nowego naboju do lufy przy równoczesnem zaryglowaniu ostatniej.

Ogólny widok pistoletu i jego części uwidoczniony jest na załączonych rysunkach (fig. 1 — 4).

Lufa *1* posiada na swej zewnętrznej powierzchni rygle *a, a* wchodzące w odnośne wycięcia zamka *2* oraz hak *b*, który, współdziałając ze ześlizgami gniazda *c*, wykonanego w chwycie *3*, powoduje podczas działania pistoletu odryglowywanie i zaryglowywanie lufy w sposób następujący: po daniu strzału, gdy zamek *2* odrzucony wtył ciągnie za sobą za pośrednictwem rygli *a, a* lufę *1*, jej hak *b* swą dolną powierzchnią ślizga się po progu *k* chwytu *3* (fig. 1); lufa w tej części drogi zamka odbywa prostolinijny ruch wsteczny, w dalszej drodze hak *b*

(Fig. 15-1) Polish Patent document of April 8, 1932

swą klinową częścią natrafia na ześlizgi gniazda *c*, wskutek czego tył lufy *1* dzięki podcięciu na dolnej powierzchni haka *b* ma możność opaść ku dołowi, przyczem rygle *a*, *a* zostają wyłączone, a lufa opierając się hakiem *b* o zderzak *L* chwytu *3* zostaje unieruchomiona, zamek zaś odbywa dalszy ruch wsteczny aż do zderzenia się jego krawędzi *m* z progiem *n* chwytu *3*.

Po ukończonym odrzucie zamek *2* pod działaniem sprężyny *23* wraca ku przodowi, ustawiając swym czółkiem *o* lufę *1* w normalnem położeniu.

Jak widać z rysunków i sposobu działania, powierzchnie haka *b* lufy *1* oraz ześlizgów gniazda *c* chwytu *3* składają się z płaszczyzn.

Zamek *2* zawiera prócz ogólnie znanych części jak: iglica *4*, sprężyna igliczna *5*, zasuwka igliczna *6*, muszka *7*, celownik *8*, również specjalnego kształtu pierścień *9* (fig. 4), który, służąc za łożysko lufy *1*, unieruchomia zamek przy jego tylnem położeniu przez oparcie się krawędzi *d* pierścienia *9* o przedni rygiel *a* lufy *1*.

Z lewej strony zamek *2* posiada wykroję *e*, *f*, *g*. Wykrój *e* służy do samoczynnego ryglowania zamka w tylnem położeniu po wystrzeleniu ostatniego naboju (zmiana magazynka) zapomocą występu *h* zatrzasku zamkowego *10*. Wykrój *f* służy do unieruchomienia zamka przy rozbieraniu pistoletu przez zaczepienie go zębem *i* bezpiecznika dźwigniowego *11*. Wykrój *g* służy do zatrzymania zamka lekko cofniętego ku tyłowi zapomocą bezpiecznika *11* w położeniu, przy którem przy spuszczeniu kurka następuje uderzenie tego ostatniego nie w iglicę, lecz w krawędź *j* zasuwki iglicy *6* (bezpieczne spuszczenie kurka).

Chwyt *3* zawiera normalne części składowe pistoletów samoczynnych, a więc: kurek *12*, spust *13*, zaczep kurkowy *14*, przerywacz *15*, bezpiecznik samoczynny *16*, bezpiecznik dźwigniowy *11*, wkładkę chwytu *17*, sprężynę kurka *18*, drążek kurka *19*, sprężynę mechanizmu spustowego *20*, zatrzask magazynka *21*, okładki rękojeści *22* i t. d.

Sprężyna główna *23* jest osadzona na trzonie *24* i opiera się swoim końcem o główkę *25* trzonu przytrzymywaną zapomocą przetyczki *26*. Pierścień *27* osadzony jest luźno na trzonie *24* i służy do przenoszenia ciśnienia sprężyny *23* na zamek *2*. W ten sposób części składowe *23*, *24*, *25*, *26* i *27* stanowią jeden zespół przy rozbieraniu i składaniu pistoletu. Zatrzask zamkowy *10* prócz ryglowania zamka w tylnem położeniu po wystrzeleniu ostatniego naboju, jak to było wskazane wyżej, służy również do związania w jedną całość pistoletu rozkładanego przy czyszczeniu na lufę *1*, zamek *2*, chwyt *3*, sprężynę główną *23*, (wraz z częściami *24*, *25*, *26* i *27*).

Rozbieranie pistoletu odbywa się w sposób następujący: po wyjęciu magazynka *28* (fig. 1 — 3) należy zamek *2* cofnąć wtył, unieruchomić go dźwigniowym bezpiecznikiem *11*, poczem pociągnąć trzon *24* ku przodowi i wyjąć zatrzask zamka *10*; po zwolnieniu zamka *2* z zęba bezpiecznika *11* zamek *2* bez przeszkód zsuwa się z chwytu *3*.

Składanie pistoletu odbywa się w kolejności odwrotnej, a więc po wprowadzeniu zamka *2* z umieszczoną w nim lufą *1* i sprężyną główną *23* na chwyt *3*, należy unieruchomić zamek zapomocą bezpiecznika dźwigniowego *11* i, naciągając trzon *24* ku przodowi, założyć zatrzask zamkowy; po zwolnieniu z bezpiecznika *11* zamek *2* pod działaniem sprężyny samoczynnie wraca ku przodowi.

Całkowite rozebranie pistoletu na części składowe odbywa się po wyjęciu bezpiecznika dźwigniowego *11*, którego ramiączko *p* przytrzymuje ośki *29* i *30*.

Zastrzeżenia patentowe.

1. Pistolet samoczynny, znamienny

tem, że odryglowywanie, zatrzymywanie i zaryglowywanie lufy odbywa się zapomocą bezpośredniego współdziałania haka, znajdującego się na jej zewnętrznej powierzchni, i ześlizgów gniazda, wykonanego w chwycie pistoletu.

2. Pistolet samoczynny według zastrz. 1, znamienny tem, że powierzchnie haka lufy i ześlizgów gniazda chwytu, powodujące odryglowywanie i zaryglowywanie lufy, posiadają kształt płaszczyzn.

3. Pistolet samoczynny według zastrz. 1 ze sprężyną główną osadzoną na trzonie, znamienny tem, że główka tego trzonu jest przymocowana zapomocą przetyczki.

4. Pistolet samoczynny według zastrz: 1, znamienny tem, że jego sprężyna główna wywiera ciśnienie na zamek zapomocą pierścienia, ślizgającego się podczas działania pistoletu po trzonie sprężyny.

5. Pistolet samoczynny według zastrz. 1, znamienny tem, że przednia część lufy jest osadzona w łożysku wykonanem w kształcie pierścienia i wkręconem w przednią część zamka, który unieruchomia zamek.

6. Pistolet samoczynny według zastrz. 1, znamienny tem, że zamek jego daje się unieruchomiać dźwigniowym bezpiecznikiem w położeniu, przy którem spuszczenie kurka powoduje uderzenie tego ostatniego nie w iglicę, lecz w tylną krawędź zamka, przyczem po zwolnieniu z bezpiecznika zamek samoczynnie wraca do normalnego położenia, kurek zaś, ześlizgując się po tylnej ściance zamka, opiera się na iglicy bez spowodowania strzału.

Piotr Wilniewczyc.
Jan Skrzypiński.

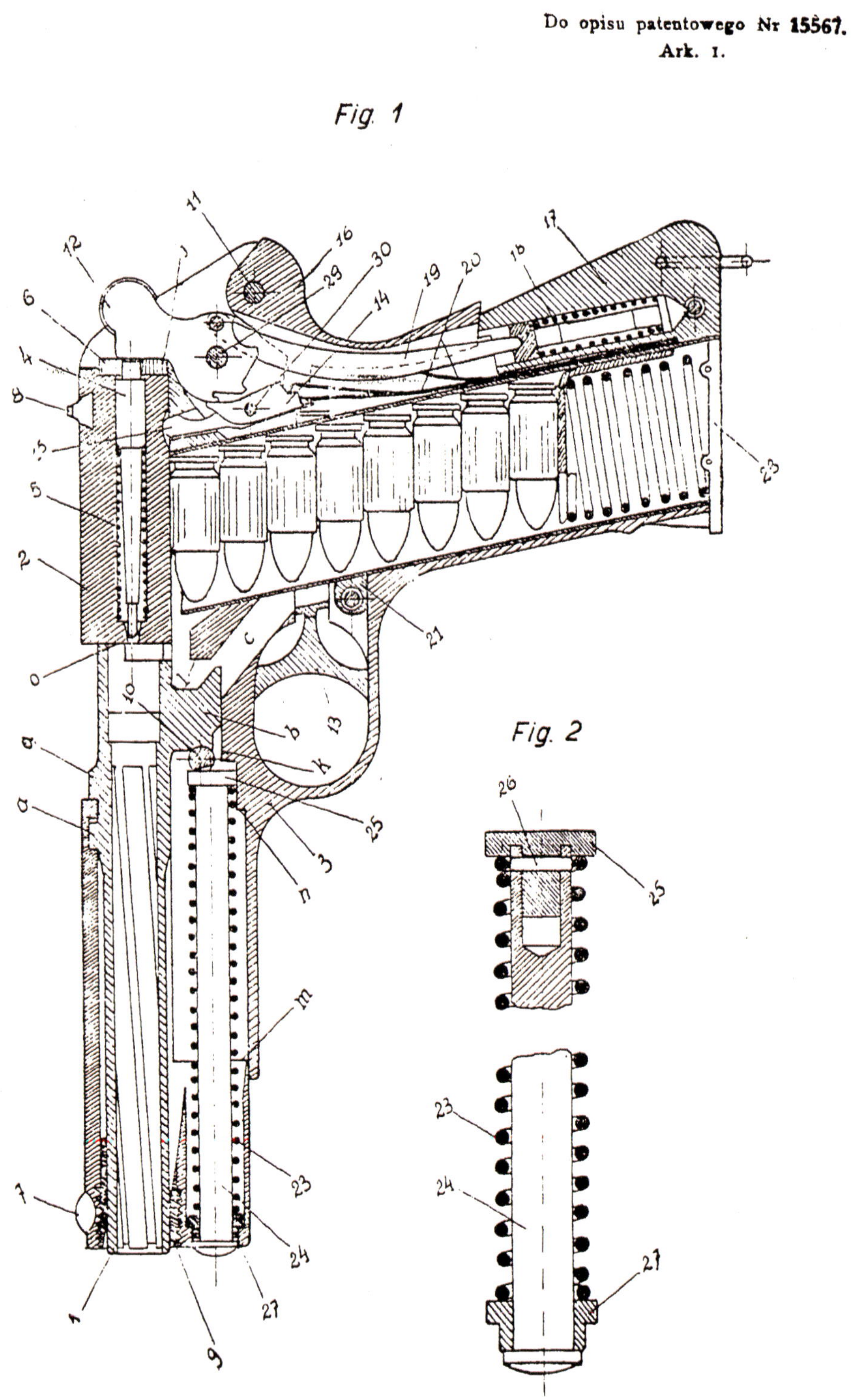

(Fig. 18-1) This patent drawing shows the captive recoil spring with its stepped ring at the front, as well as the cam locking and unlocking system.

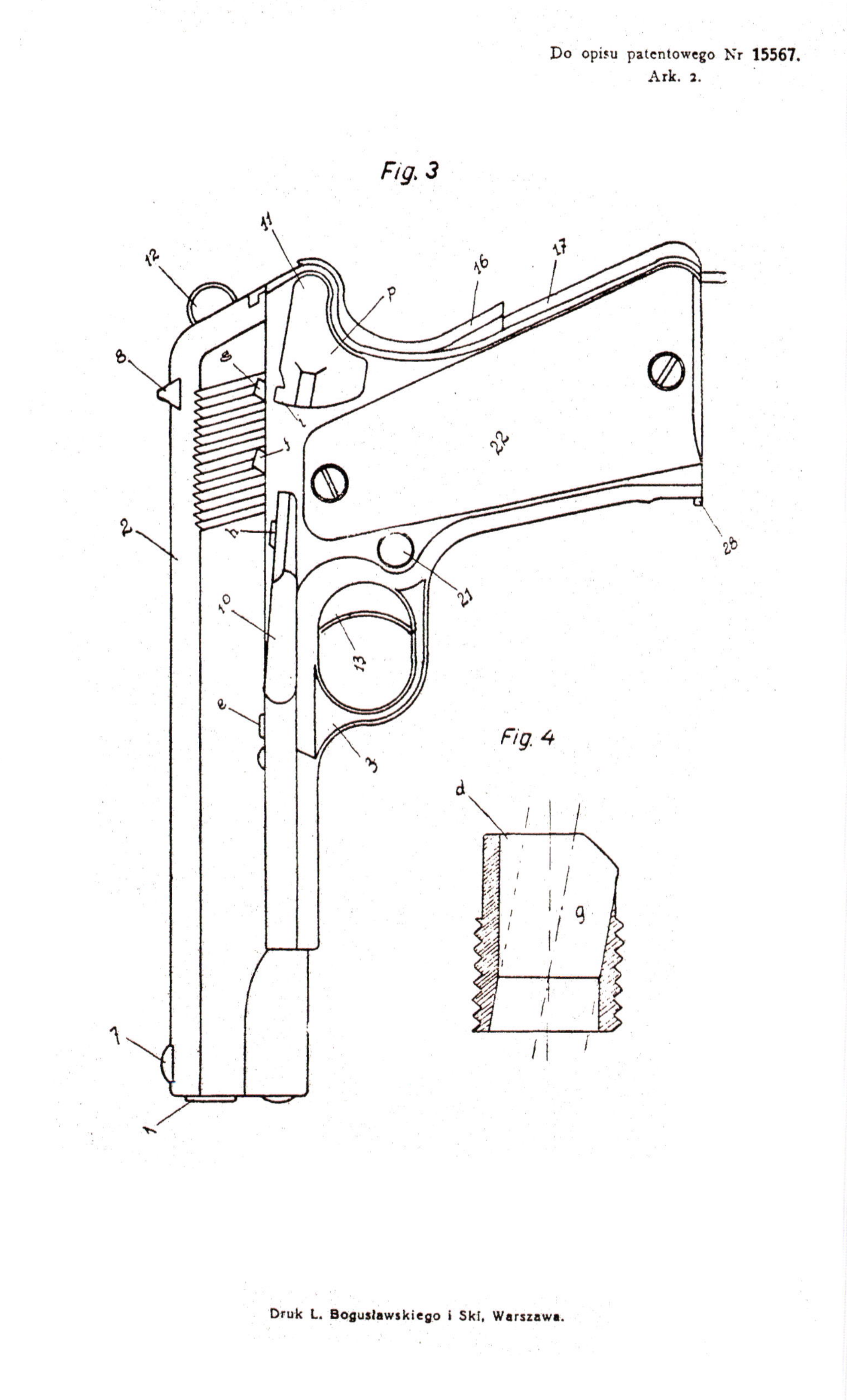

(Fig. 19-1) This patent drawing shows a screwed-in barrel bushing, and the use of the takedown lever as a safety.

Design influences: The Colt M1911

Piotr Wilniewczyc wrote in his article that rather than reinvent the wheel, he decided to use the best of the prior art in his pistol design. Since patents on Colt's M1911 pistol had expired, its design was heavily borrowed upon. A detailed side-by-side examination of the two pistols will confirm this, as will be seen in the following pages. However, the VIS differs from the M1911 Colt-Browning design in significant fashion. While the barrels of both arms are locked to their respective slides by two locking lugs on top of the barrel, mated to grooves on the upper interior of the slide, the means of locking and unlocking are quite different. In the Colt, a swinging link effects locking and unlocking. In the VIS, this is done by camming surfaces on the bottom lug of the barrel, and in the frame. Present-day Radom factory literature attributes the design of this camming system to Feliks Modzelewski. Furthermore, while the Colt's recoil spring is guided by a short rod to the rear and a cup to the front, the recoil spring in the VIS design is held captive on a full-length rod, and is removed as a unit when the pistol is stripped. Also, while the Colt's barrel bushing in the front of the slide is a separate part which must be removed in order to remove the barrel, the VIS has a permanently attached barrel bushing, the barrel being removable from underneath the slide.

At this juncture, it is interesting to examine the claims of the VIS patent of 1932, to determine just what the pistol's designers thought was new, and thus patentable, in their design. There are, in fact, six claims made in their patent, with the usual redundancy common to all patent documents. First, the system of barrel-frame interaction by camming surfaces is claimed as an invention. Second, the fact that these mating camming surfaces are flat, is claimed. The third claim is for the pinned construction of the (at that time inflexible) captive recoil spring. Fourth is for the employment of a sliding ring at the front end of the captive recoil spring. Fifth is for the permanently installed, "screwed in" barrel bushing in the front end of the slide. Sixth and last is a claim for the use of the VIS dismounting lever as a safety catch whereby it could be turned up into a second slide notch, holding the slide slightly out of battery and allowing the hammer to be lowered (or to fall) without delivering a blow to the firing pin. This last claim would also require a lengthening, or a doubling-up, of the disconnector recess on the bottom of the slide. This would seem to be inadvisable; at any rate, this claim appears to never have been made use of. In conclusion, the designers' stated innovations are:

1- Cam locking and unlocking by means of flat camming surfaces.
2- Sliding ring at front of captive recoil spring.

Design influences: The claimed Fabrique Nationale influence

Over the years since its development, numerous gun writers and firearms authorities have looked at these three claimed innovations and reached the conclusion that they are common to both the VIS and the Browning High Power or its forerunner the High Efficiency. Their conclusion often was that the VIS was designed and produced with the assistance of engineers from the Belgian firm Fabrique Nationale d'Armes de Guerre (FN). This conclusion was first speculated by Gerhard Bock on p.228 of his MODERNE FAUSTFEUERWAFFEN UND IHR GEBRAUCH (J. Neumann, Neudamm, 1941), and was cited , and attributed to a "noted German authority" by W.H.B. Smith in an article entitled "Polish Radom Pistol", part of a series entitled "This Pistol Packin' War", on pp. 19-20 of the November, 1944 American Rifleman magazine. Smith treated Bock's speculations as fact, and his conclusions have been accepted as such by many firearms writers ever since.

(Fig. 21-1) The Polish legation of engineers and engineering students pose for a group photo on the courtyard at Fabrique Nationale in 1929. They were present for the development and production of the Polish BAR variant, the wz.28. The co-designer of the VIS pistol, Jan Skrzypinski, is assumed to be in this photograph as he was present at FN during that time, yet members are not individually identified. (Photograph courtesy of Anthony Vanderlinden)

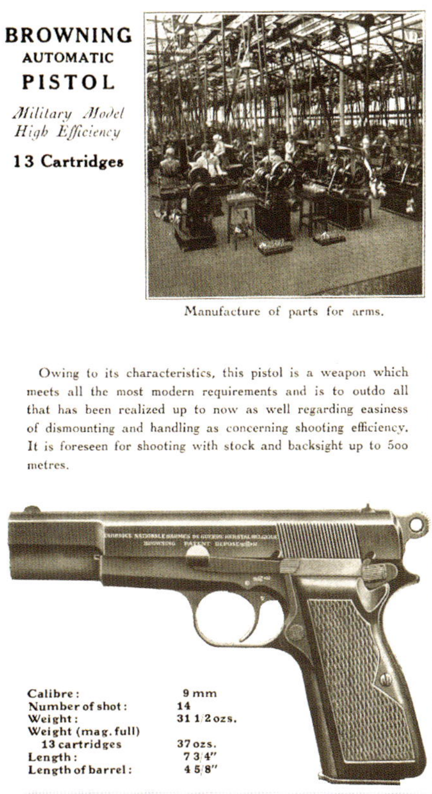

BROWNING
AUTOMATIC
PISTOL
Military Model
High Efficiency
13 Cartridges

Manufacture of parts for arms.

Owing to its characteristics, this pistol is a weapon which meets all the most modern requirements and is to outdo all that has been realized up to now as well regarding easiness of dismounting and handling as concerning shooting efficiency. It is foreseen for shooting with stock and backsight up to 500 metres.

Calibre:	9 mm
Number of shot:	14
Weight:	31 1/2 ozs.
Weight (mag. full) 13 cartridges	37 ozs.
Length:	7 3/4"
Length of barrel:	4 5/8"

(Fig. 21-2) FN's 1930 catalog announces the Grand Rendement (High Efficiency), this design variant did not have a barrel bushing. (Catalog page courtesy of Anthony Vanderlinden)

Recently, however, there have been some rumblings to the contrary. Robert J. Berger, who was perhaps the first person to publish an English language translation of both the Wilniewczyc article and the Polish patent, feels that Bock's speculation is wrong. Peter G. Kokalis, in his article on the VIS in Shotgun News, agrees with Mr. Berger that the theory is a myth. With this in mind, let us look at the actual arms in question, as well as some others, and try to draw some conclusions.

Polish records of small arms reveal that at least some "Browning M 1930" pistols were present in Poland (in fact, but for financial considerations, the Polish government nearly adopted the model), and we must assume that the characteristics of these pistols were known to the VIS designers. The ordnance engineer and director of the State Rifle Factory in Warsaw, Jan Skrzypinski, spent much time at Fabrique Nationale in 1929 with the development and production of the Polish Browning Automatic Rifle. This time frame coincides with the public announcement and release of the Grand Rendement. FN, which had been developing the pistol for many years was eager for a military contract and the Polish delegation was without doubt introduced to the new pistol designs.

The "Browning M 1930" referred to in Poland was one of the evolutionary precursors of the 1935 Browning High Power, known and marketed at the time as the High Efficiency.

The "M1930" designation was a Polish designation and it is not clear if the Polish designers were looking at the Fabrique Nationale design variant with removable barrel bushing, or the 1929 variant without barrel bushing, or both. As such, it would indeed have had a cam-locking system, with its barrel cam on the underlug. However, the High Efficiency and the later 1935 High Power have a rather long recoil spring guide, but not a captive recoil spring like that in the VIS.

Let us now more closely compare the barrel lug camming surfaces of the Browning and of the VIS. At a glance, it will be seen that they are not at all the same. The camming surfaces used on the Browning design are internal to the barrel lug, while those on the VIS are external. Actually, the VIS lugs appear to be easier to machine. The VIS patent of 1932 claimed the invention of the permanently attached barrel bushing, a feature which was to appear in the 1930 Browning and first mass produced by FN in 1933 for Belgian military trials. Yet, it is doubtful that there were Polish engineers assisting the Belgians. Rather, these would seem to be examples of parallel development.

(Fig. 22-1) The barrel and recoil spring of a 1933 FN Browning High Efficiency. (Photograph courtesy of Anthony Vanderlinden)

Design influences: The .45 caliber Ruby

Moving away from the Browning to other possible design sources, let us begin with the captive recoil spring. Pistols so equipped were in fact quite common in 1920s Poland. They were the principally Spanish-produced Ruby, known as the Cebra in Polish, which had been produced in great numbers for French and Italian forces during the First World War. The Polish military published a manual for this weapon in 1922, the parts diagram clearly showing its captive recoil spring. The VIS designers were certainly aware of the Ruby's features, but we should keep in mind that the standard Ruby was of small caliber (usually 7.65mm) and less power. Also the final VIS design of a "flexible" recoil spring rod is quite different from either the Ruby's simple captive spring, or from the VIS patent claim of 1932.

However, there was another Ruby on the scene; the little-known Ruby .45. Gabilondo y Cia. began to make locked breech pistols in .45ACP in 1924 and sold them under the "Ruby" trade name. Ignacio Ugartechia made them under the trade name "Inaki". Production was limited as the firms soon transitioned to pistols that resembled the Colt M1911.

This arm is hard to find today, as apparently fewer than 1,000 examples were ever produced. A stripped example was first illustrated in Dr. Edward C. Ezell's HANDGUNS OF THE WORLD. The pistol has a removable barrel bushing, a captive recoil spring assembly, and an external underlug cam locking system which much resembles that used on the VIS. The rest of this pistol appears to be a simplified version of the basic Browning design of the Vest Pockets like the FN Browning 1905 (1906) and the Colt Model 1908. Yet, it presents us with some intriguing questions. Was this pistol the real source of the VIS cam locking design? The illustration we have seems to show some slightly curved surfaces on its barrel lug. Is that why the Polish patent specifically claims flat camming surfaces? We will probably never be certain that this pistol was known to the Polish designers, but it probably was. It always may simply be another case of parallel development. Notably, the Ruby .45 also has a sliding ring at the front of its captive recoil spring, albeit a plain, unstepped one.

(Fig. 23-1) The Ruby .45 first appeared in Gabilondo y Cia's N°25 catalog that was presumably printed in 1927. Additionally recorded serial numbers include N° 221, 415, 437, 547, 550, and 642. A few guns were sold to Ignacio Ugartechea who commercialized the pistol under the name IÑAKI. Production likely ended in 1931 upon introduction of the "Llama" marked .45ACP whose design was more closely based on the Colt M1911. (Photographs courtesy of Leonardo M. Antaris, MD)

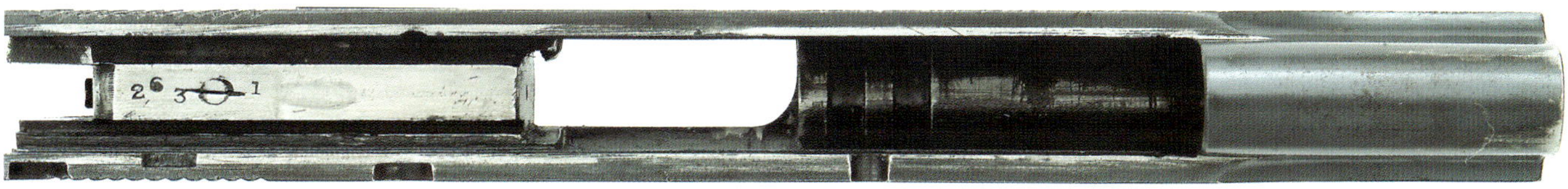

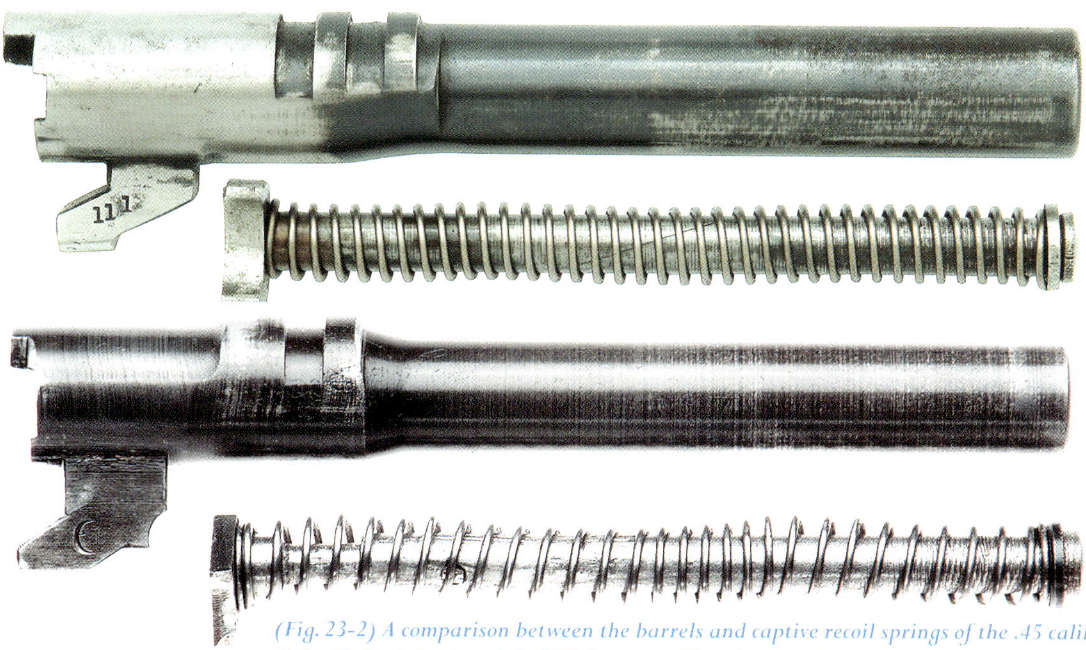

(Fig. 23-2) A comparison between the barrels and captive recoil springs of the .45 caliber Ruby (Cebra) (top) and the VIS (bottom). The photographs are not to scale. (Ruby photographs courtesy of Leanardo M. Antaris, MD)

(Fig. 24-1) Right side view of the Ruby .45
(Photographs courtesy of Leanardo M. Antaris, MD)

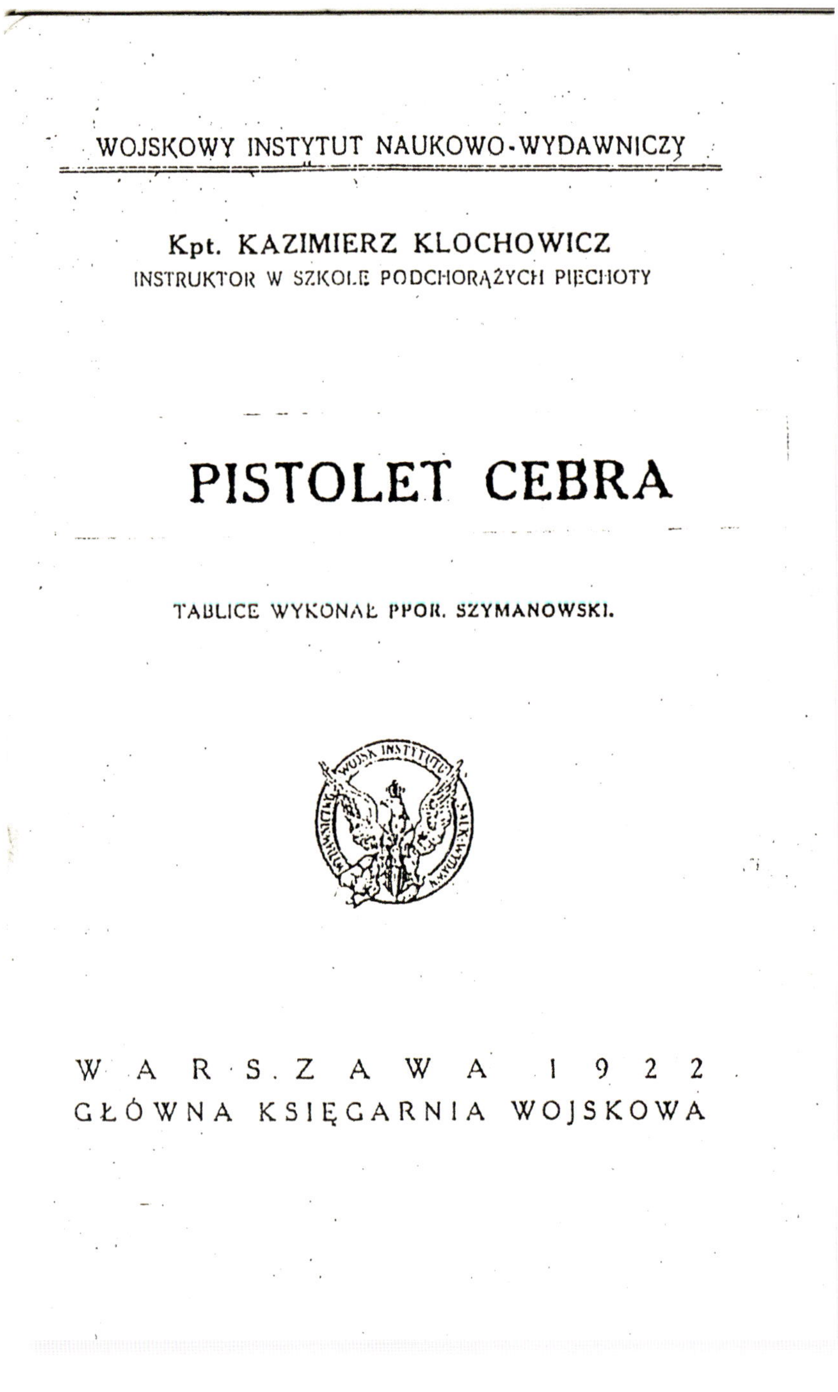

WOJSKOWY INSTYTUT NAUKOWO-WYDAWNICZY

Kpt. KAZIMIERZ KLOCHOWICZ
INSTRUKTOR W SZKOLE PODCHORĄŻYCH PIECHOTY

PISTOLET CEBRA

TABLICE WYKONAŁ PPOR. SZYMANOWSKI.

W A R S Z A W A 1 9 2 2
GŁÓWNA KSIĘGARNIA WOJSKOWA

(Fig. 24-2) Polish manual for .32 Cebra
(Ruby) dated 1922 (Author's Collection)

Chapter III
The VIS Pistol - A Detailed Parts Description

The VIS is a single action semi-automatic pistol of the Browning type, with an external hammer and a barrel-to-slide locked breech mechanism. It has a grip safety, a hammer drop lever, and a "floating" type firing pin. Its removable magazine is of single-column type, and holds eight cartridges. Its caliber is 9x19 mm Parabellum (Luger). The pistol's main subassemblies, not counting the magazine, are the barrel-slide assembly and the frame assembly. We shall now examine each of these.

Barrel-Slide Assembly

This assembly consists of the barrel, the captive recoil spring (itself an assembly), and the slide assembly proper. It is attached to the frame assembly by the slide stop, of which more will be said later.

The **BARREL** is chambered for the 9x19mm cartridge, and is rifled with six grooves, right hand twist. On its lower breech end is an integral lug, which will usually be found to be stamped with a serial number, a firing proof, and assemblers' or sub-inspectors' marks. Polish prewar guns will also ordinarily have a metric bore diameter stamped on the lug. The Germans moved this mark to the barrel in front of the lug. The lower part of this lug is machined to form camming surfaces which interact with corresponding frame surfaces to raise or lower the rear part of the barrel, thus effecting the locking or unlocking together of barrel and slide. On the upper rear surface of the barrel, opposite the barrel lug, are two raised and machined locking lugs, which are fitted to corresponding locking grooves in the slide and hold slide and barrel fast together at the moment of firing.

(Fig. 25-1) The metric bore diameter marking on a prewar (left) and wartime (right) barrel. The Germans moved the marking from the barrel lug to the barrel body. (Photographs courtesy of Anthony Vanderlinden)

The **CAPTIVE RECOIL SPRING ASSEMBLY** is removed from the pistol in one piece, and should not be disassembled further. In prototype versions of the VIS, and in late wartime examples, it consists of a simple rod with an enlarged front end, on which are fitted a sliding, dual diameter washer at the front end, the spring itself, and a back piece which interacts with the slide stop and barrel lug. However, in its usually encountered form this assembly is complicated by the addition of an inner coil spring of small diameter inside a hollow spring rod. This internal spring works on a small diameter rod, attached as a sort of "tail" to the assembly's back piece. This "tail" is laterally slotted and, to an extent, flattened, to allow it to be held by a retaining pin.

This pin is inserted through a hole in the side of the hollow spring rod, and enters through an enlarged opening at the rear of the slotted inner rod or "tail". This pin is quite small, and is shaped rather like a dumbbell with reduced diameter in its middle, so it may run in its slot. As was stated above, the purpose of this assembly is to soften an unexpected blow on the front of the pistol, as from a fall, and to allow the barrel to slip out of battery, thus preventing an accidental discharge. This assembly is quite cleverly conceived, but it is delicate and easily put out of order by rough handling or an improper attempt at disassembly. It was reluctantly added during the pistol's initial testing by Piotr Wilniewczyc, who viewed such an accidental discharge as a remote possibility.

(Fig. 26-1) Captive recoil spring assemblies. Comparison of solid and articulated, with parts of the articulated assembly shown. (Photograph courtesy of Anthony Vanderlinden)

The **SLIDE ASSEMBLY PROPER** consists of the slide body, which is machined to accept the barrel and lock it into place, the permanently attached barrel bushing, the decocking lever, the firing pin, spring, and stop (or retainer) the extractor, and the rear sight.

The top of the slide includes an integrally milled front sight and a matted sighting rib, which runs from front sight to rear sight on early pistols, and to the end of the slide on later wartime pistols. A dovetail cut for the rear sight is also present.

The front of the slide is machined to accept the permanently mounted barrel bushing, which is shown and stated to be threaded in the 1932 patent, and is held in place by a lateral pin. An uninstalled VIS barrel bushing in the author's possession is not threaded, which leads him to speculate that production, or at least wartime, bushings were simply press fitted in place and pinned.

(Fig. 26-2) At the left is an uninstalled VIS barrel bushing, at the right is a wartime VIS slide which has been skeletonized at the front to show the pressed-in and pinned barrel bushing. (Photograph by Anthony Vanderlinden, author's collection)

The slide bottom is machined to accept the barrel, to allow the barrel to be locked in place, and for installation on the frame. Its rear area is slotted for the ejector, and includes a disconnector recess, which allows the pistol to fire only when fully locked- when it is "in battery". This area of the slide also includes a cut to enable the internal parts of the decocking lever to cam down the sear-disconnector and drop the hammer while also camming the firing pin into the slide where the hammer cannot deal it a blow. The slide is also machined to accept the base of a chambered cartridge, and includes holes for firing pin and extractor. The slide bottom is extended downward on both sides, and is machined to mate with, and stay attached to, the frame.

The rear of the slide is bored out for the firing pin and its spring, and for the extractor. It is machined to accommodate the firing pin stop, which also retains the extractor and the hammer drop lever.

The left side of the slide is, in earlier guns, slotted for the hook of the dismounting lever, and provided with a semicircular cut which allows the removal of the slide stop during disassembly. Later slides lack the dismounting lever cut and have the slide stop cut repositioned forward on the slide.

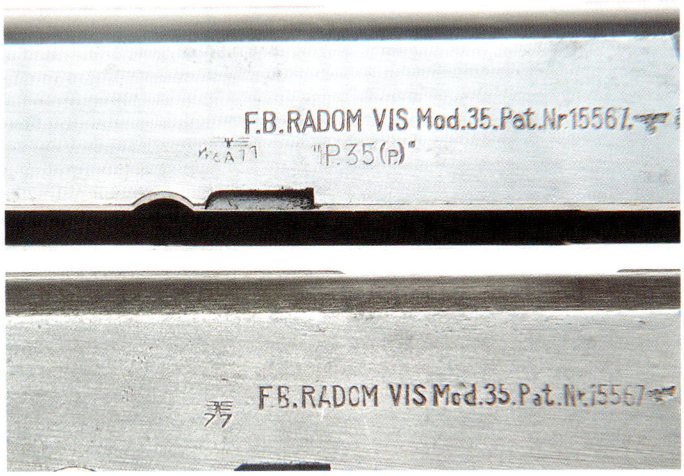

(Fig. 27-1) Top: An early wartime VIS slide showing the position of the dismounting cut. Bottom: A late wartime VIS slide with its dismounting cut repositioned forward. (Photograph courtesy of Anthony Vanderlinden)

The hammer drop lever is also mounted on the left side of the slide. It is returned to its position by a small piano wire spring, and has internal cams as described above. The lever itself is fabricated of two parts, the attachment of which is often visible toward the rear of the lever. This is another assembly the construction of which shows considerable genius, and which was added to the VIS, apparently at the behest of the cavalry, during the testing stage. The takedown lever is also fabricated of two parts. It is common to find VIS pistols with crude repairs where the joints in these levers have become loose. (see fig.211-1)

Both sides of the slide are provided with grasping grooves toward the rear, as an aid to manually cycle the slide.

The firing pin stop, or retainer, is installed at the rear of the slide , usually quite tightly and often "dead tight". It should not be routinely removed, especially in later war pistols. Late war firing pin stops are sometimes brittle and may crack if care is not taken. This part is not a simple plate, as in the Colt, and is machined internally for the hammer drop lever.

The VIS firing pin has a long, thin front section which is subject to breakage. Its rear section is sturdy, and has a flat surface machined on it to accommodate the hammer drop's camming surface. The firing pin is of the "flying" or "floating" rather than the "full reach" type. Its coil return spring is the same as the M1911 Colt's.

The extractor is a long, spring tempered piece with a hook at its front end for the cartridge rim and a slot toward the rear so it may be held in place by the firing pin retainer. VIS extractors were individually fitted and were made "too long" so they could be ground to final contour with the slide. Replacement extractors will seldom fit a slide properly.

The Slide Stop

This is the part that keeps the slide and frame together on the VIS. It is made in one piece, and consists of a thumb lever and a pin. The area under the thumb lever is so shaped as to allow the magazine follower to raise the stop and hold the slide open when the magazine is empty. The slide stop pin is flattened on one side, and interacts with the back-piece of the captive recoil spring, which holds the slide stop in the pistol, and which also serves to hold the stop in its lower position when it is not being acted upon by the magazine follower. The other side of the pin is rounded, and bears against the barrel lug, holding the barrel in its locked position when the slide is forward.

The Frame

The frame of the VIS is very similar to that of the Colt M1911 pistol. It is made with an integral trigger guard, and has rails machined along its top sides to accommodate the slide. Its grip or handle area is milled out for the magazine, and the frame is laterally holed for the magazine catch, slide stop, dismounting lever (on later production guns, for the grip safety pin), hammer, sear-disconnector, ejector, and mainspring housing pins, and for the grip screws or bushings. The rear of the frame is hollowed out for the trigger, sear, disconnector, multipurpose spring, grip safety, hammer, and mainspring housing, with grooves at the bottom to retain the mainspring housing. The frame top is holed for the disconnector, and slotted for the thin, flat ejector, which is held in place by a thin pin. The top of the frame is also hollowed out in front for the recoil spring, and the back part of this area is provided with camming surfaces which interact with the barrel lug's cams to enable locking and unlocking.

The sides of the frame are also cut out under the grips, and, as we just noted, are holed, drilled and tapped for grip screw bushings, or for the enlarged late war screws themselves (which are the same diameter and thread as the earlier-war bushings).

The magazine catch is inserted from the right side of the pistol, behind the trigger guard, and is retained in almost all instances by a small, screwdriver slotted, retainer, backed by a coil spring. The spring powers the catch in its job of magazine retention. The catch also retains the trigger in the frame.

The dismounting lever is mounted at the rear of the frame, its pin retaining the grip safety. While this lever resembles the safety catch of the Colt M1911 pistol, it is in no way a safety catch. Its only purpose is to hold the slide in its proper position for dismounting. The outside "plate" of this lever retains the hammer and sear-disconnector pins. The lever's own pin has a small lug on its side which keeps the pin in the frame, and the lever can only be inserted or removed by rotating the lever vertically beyond the top of the frame prior to pulling the lever out.

The grips retain the mainspring housing pin, and are retained by four screws, two on each side. Grips on early pistols are made of hard rubber, while later wartime grips are made of black or brown Bakelite, or of wood.

The mainspring housing is secured at the bottom by a headless pin of the same diameter as the hammer pin. The housing is hollow and contains the mainspring and its cap, both retained by a thin pin inserted laterally. The mainspring itself powers the hammer, while the housing has the secondary task of retaining the multipurpose sear-disconnector, trigger, and grip safety spring. The latter is similar, but not identical, to that in the Colt M1911 pistol.

The hammer is held in the frame by a lateral, headed pin. It is itself pinned to the hammer strut, which bears against the mainspring cap when the pistol is assembled and transmits the mainspring's energy to the hammer.

The sear and disconnector are held together, and retained in the frame, by a thin lateral headed pin. These parts are adapted to the VIS hammer drop system, and thus differ in form from the corresponding parts in the Colt M1911.

The VIS trigger is similar to that in the Colt M1911, in having a rectangular bow which runs in a cut milled into both sides of the magazine well. The rear of this bow is shaped to accommodate the sear and disconnector. Early triggers were milled from a single piece of steel; later wartime ones are fabricated, with rivets visible on both sides of the finger piece.

The grip safety is modeled after that in the Colt M1911. When the pistol is grasped properly, the safety allows the pistol to fire. When the safety is not depressed, its internal lug prevents the motion of trigger and sear, preventing the pistol from firing.

VIS pistols manufactured late in World War II lack the takedown lever, as this was deleted for the purpose of simplification. This necessitated two other changes. First, the hammer was changed by adding a hook to its upper surface so it could engage the internal hammer drop lever. Second, the takedown cut on the slide was moved in order to allow takedown with the new arrangement. Also, the lateral pins holding the hammer, grip safety, and sear-disconnector, were cupped at both ends and permanently riveted in place. These pins are often referred to as "rolled" pins, but they are not. They are solid, cupped pins, upset (riveted) in place.

(Fig. 29-1) Left: View of dismounting hook on top of late-war VIS hammer
Right: Late-war slide held in takedown position by hammer drop lever and hammer hook. (Photographs courtesy of A.Vanderlinden)

The Magazine

The VIS magazine is of single column construction and is of basic Browning type. It consists of a magazine body, spring, follower, bottom plate, and two bottom plate pins. The followers of most VIS magazines are of cupped and machined construction, and are so shaped as to bear on the slide latch and hold the pistol open when empty.

Polish and early German produced VIS magazines are essentially identical with the exception of their markings. These magazines have extruded bodies, formed followers, and pinned bottom plates. Most Polish produced magazines have the oval letter-and-number acceptance mark at the toe, although some late ones are unmarked.

German production magazines are mostly unmarked save for a workman's or sub-inspector's mark on the tube. The quality of finish of these wartime magazines will be seen to decline as the war progressed. Very late war followers are made of pressed sheet steel, so shaped as to work the slide latch. The latter are often referred to as "P38" followers, but they are actually quite different, and specific to the VIS magazine. Wartime magazines are often found with the (E/623) acceptance marks either on the tube or on the bottom plate (E/189). Final German production magazines are phosphate finished and are fitted with stamped followers.

The bottom plates of most VIS magazines are pinned to the magazine body and should not be removed unless this cannot be avoided, as in the case of a badly dented magazine that must be straightened by use of a mandrel. Later war magazine bottoms appear to be crimped to the body rather than pinned, and are likely even more of a problem to dismount. The earlier cupped followers cannot be removed from the top without severe damage to the magazine. The distinctive late war magazines with stamped followers are quite rare, so such removal should not be attempted.

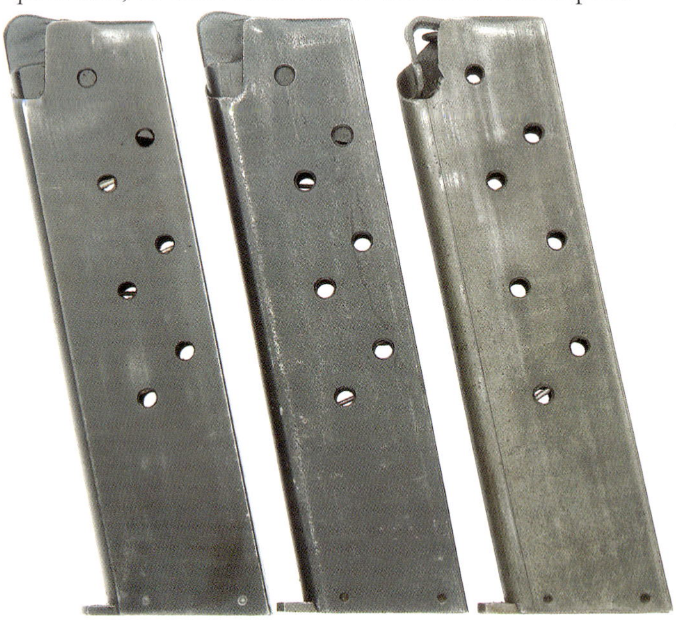

(Fig. 30-1) Prewar (blue), early (blue), and late war (phosphate) magazines. Many wartime and some prewar magazines are unmarked. The unmarked wartime magazines are assumed to have been made at Radom. The late war phosphate-finish magazine illustrated has the so-called "P38" type follower. It is actually a unique stamped follower similar to, but not compatible with, the P38 magazine's follower. (Photograph by Anthony Vanderlinden, courtesy Waclaw Ustupski collection)

(Fig. 30-2) Typical prewar magazines have the Polish final inspection marking on the toe. (Photograph by Anthony Vanderlinden)

(Fig. 30-3) Magazines made during the war at a Steyr subplant in Warsaw are E/189 marked, magazines made at the Radom plant were unmarked. (Photograph by Anthony Vanderlinden)

(Fig. 30-4) Late war phosphate magazines were made at Steyr or Gusen and E/623 marked on the body. (Photograph by Anthony Vanderlinden)

Grips

The grips on prewar made VIS pistols are made of black hard rubber. They are checkered, and display the factory's initials "FB" on the left side, and the pistol's name "VIS" on the right, both within triangles. German occupation era grips may be of black hard rubber (early), or of black or brown Bakelite (later), with identical markings, or of pressed wood. Wooden grips with pressed-in identical markings may be found. The last group of grips on German production are Ersatz grips of grooved wood.

Wooden or Bakelite grips which are checkered and otherwise unmarked, are considered Ersatz field replacement grips.

Late production German grips have large screw holes to accommodate the later, bushing sized grip screws. Late war Steyr grooved grips are typically E/623 proofed on the back side, but the marking is small and difficult to observe.

(Fig. 31-1) Prewar and early wartime black grips have no visual differences, 2) wartime brown or black Bakelite grips, 3) wartime pressed wood grips, 4) late war Steyr grooved wooden grips.(Photograph by Anthony Vanderlinden, courtesy Waclaw Ustupski and Francis Kennedy collections)

(Fig. 31-2) Wartime field Ersatz grips come in many levels of finishing: Seen here are fine and coarse checkering. Bakelite field Ersatz grips do exist as well. (Photograph by Anthony Vanderlinden, courtesy Waclaw Ustupski collection)

(Fig. 31-3) Prewar and early wartime grips screws (left) and later wartime bushing-sized grip screws (right). (Photograph by Anthony Vanderlinden)

Chapter IV
VIS Production History and Characteristics

Prewar VIS Production

During the years 1936-1939, the Radom arms factory employed over 4,000 workers in the businesses of producing arms as well as motorcycles, bicycles and other civilian items, a mix mirroring the product variety produced at Fabrique Nationale in Belgium and at Czech and other arms factories.

The VIS pistol was manufactured by traditional methods of forging, machining, assembly, inspection and finishing. This is true of both prewar Polish and wartime German occupation production.
Prewar Polish production of the VIS was centered in the town of Radom, and is characterized by a high level of quality in both fit and finish. There are few visible variations in prewar pistols with the exception of the year marking and assembly and inspection marks; although a close inspection will reveal that pistols produced in 1936 have front sights with a rounded profile when viewed from the side, while the side profile of all later VIS front sights sloped up toward the muzzle. The steel used in the first 3000 VIS pistol frames made in 1936 was an alloy steel, and all later production used carbon steel; but this is not a visible difference.

The VIS pistol, while a quality product, was not manufactured to anything like the standard of interchangeability found in contemporary United States weaponry. For example, the final machining of the rear of the slide was done with the extractor in place. The rear of the frame was also machined in its final stages with the grip safety and mainspring housing in place. We have already noted that the former becomes obvious when an extractor is replaced. The individual fitting of the grip safety and mainspring housing are obvious on crude late-war guns, and hamper the fitting of replacements on all VIS pistols. Due to this individual fitting, Polish era VIS pistols carry serial numbers on nine components; frame, slide, barrel, hammer, grip safety, mainspring housing, hammer drop lever, firing pin retainer, and extractor.

(Fig. 32-1) A rare photograph of the Radom factory taken within days of the invasion. The photograph is marked in pencil "Ausgebrannte Waffenfabrik in Radom" (Burned out weapons factory in Radom). (Photograph courtesy of Anthony Vanderlinden)

The VIS pistol itself appears to have been manufactured in at least two production lines, and required much hand fitting of its major assemblies prior to final finishing.

It must be noted that approximately the last 2000 VIS pistols produced in 1939 while the Radom factory was still under Polish control were "mismatches" made of previously fitted parts put together into pistols without regard to serial number integrity, obviously during the German onslaught. These pistols also usually lack the lanyard loop at the bottom of their mainspring housings; a part which in normal times was apparently added during final assembly.

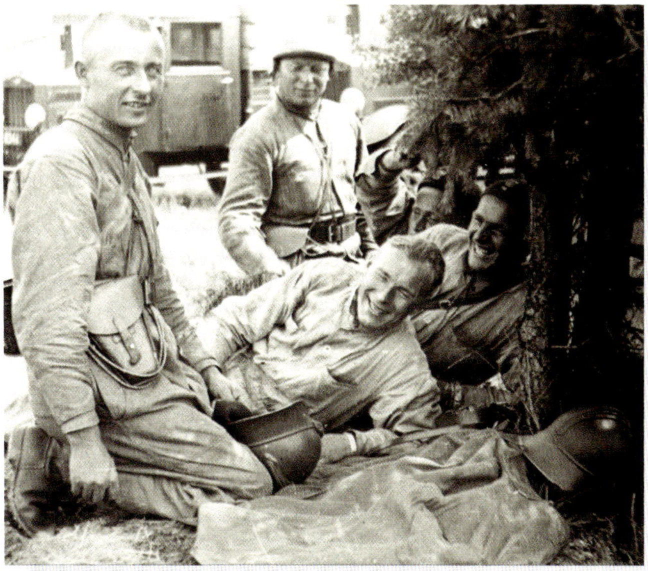

(Fig. 33-1) Various photographs of Polish soldiers and officers with their VIS holsters. (Photographs courtesy of Waclaw Ustupski)

(Fig. 34-1) Polish mechanized cavalry soldiers with VIS holsters, photograph most likely taken during the German invasion. (Photograph courtesy of Waclaw Ustupski)

(Fig. 34-2) German infantry crouches down for Polish fire.

(Fig. 34-3) Polish commanders resist the German onslaught, using all available resources including barricading city streets with trolley cars.

(Fig. 35-1) Street fighting in Warsaw; German infantry clears Polish barricades.

(Fig. 35-2) One of many gathering points for captured Polish war material.

(Fig. 36-1) Polish POWs are led away while the German Army advances.

(Fig. 37-1) A rare photograph showing lined-up captured VIS holsters. (Photograph courtesy of Waclaw Ustupski)

(Fig. 37-2) Polish officer surrendering

(Fig. 38-1) General Blaskowitz dictates the articles of surrender to Polish government representatives.

(Fig. 38-2) Hitler inspects the banner of the defeated Polish 8th Rifle Regiment.

Production during the German Occupation: Radom, Steyr & Znaim

When German forces seized the Radom factory in September, 1939, they immediately set about making an inventory of all buildings, machinery, raw materials and partially or wholly finished products on hand. They then assigned a new management staff to the factory, headed by a trustee, who was an official of the Steyr-Daimler-Puch arms factory in Austria. The Steyr works was a large arms-making consortium located a relatively short distance (about 400 miles) from Radom, and was closely connected to the Nazi regime. From the German point of view, this made Steyr the ideal manager of the Radom industrial plant. The Radom plant remained idle for a year after its inventory was secured. The reasons behind this are unknown, it has been reported that this was due to the location of Radom and the proximity to the German Soviet demarcation line. A cursory review of the demarcation line clearly shows Radom as being well into the German zone.

(Fig. 39-1) The "General Government" - Nazi German occupied Poland. Prior to the June, 1941 invasion of the Soviet Union, it did not include the Lemberg (Lwow) District.

Due to the fact that a large portion of the Polish workforce at Radom had departed from the factory and refused to work for the occupying Germans, a system of compulsory labor was introduced at the Radom plant once the plant was restarted. Eventually, the factory was to rely heavily on Jewish slave labor, with over 1000 Jewish workers who were housed in a nearby purpose-built concentration camp.

The VIS pistol was classified by the Germans as the P35(p) (P35 polnisch - Polish). Production during the German occupation began in the latter part of 1940 with the assembly of pistols from parts found in inventory. This first production phase lasted until parts supplies were exhausted (not all at the same time). These pistols may be found with prewar Polish slide legends over-stamped with the German P35(p) model designation, along with German proof and Waffenamt acceptance marks. This German production phase is said to have begun in September of 1940, with full serial production having begun by March or April of 1941.

(Fig. 40-1) Germans inspect and inventory wz.28 Browning machine-guns inside the Radom factory as part of their assessment of seized inventory. (Photograph courtesy of Anthony Vanderlinden)

Early German production kept the prewar parts numbering pattern, but the numbering of extractors and firing pin retainers was later dropped as one of many production simplifications.

As the war progressed, the pressing need for increased weapon production led to a steady downgrading of fit and finish and an overall decline in quality of wartime guns. The hard working conditions which were imposed on an underfed and unwilling workforce also were contributing factors. Design simplifications were introduced over time in order to increase output performance.

Certain of the forced workers at Radom managed to steal slides, frames and related parts which were used to produce pistols for the Polish underground. All wartime barrels were produced at Steyr, which was also the point of final wartime assembly, so underground VIS barrels had to be either scrounged or made from scratch in illegal workshops. Two such workshops are known to have operated in Warsaw, one taking over production after the Germans had raided the other. Radom workers stealing pistol parts faced the death penalty, which is reflected today by a memorial at the factory honoring those workers who were executed for these activities.

Besides the moving of barrel production and final assembly to Steyr, work on the VIS took place much as it had before the war, with the exception being that the Radom factory only produced pre-fitted VIS pistol "kits", and sent them on to Steyr for fitting with Steyr-produced barrels and final acceptance. The multi-location nature of wartime VIS production and assembly up to the end of 1944 is evidenced by the presence of German Weapons Office (Waffen Amt) acceptance marks from both locations on these pistols.

The effort of shipping "kits" to Steyr and the desire to prevent theft by the Underground eventually led to the assignment of individual inventory numbers to unfinished frames and slides; a later war characteristic.

Magazines were manufactured by three different facilities under Steyr's control (Radom, a Steyr-subplant in Warsaw and Steyr in Austria).

Over time the VIS was simplified by elimination of the takedown lever, the use of riveted frame pins, and design changes which enabled the hammer drop lever to do double duty as a takedown lever. Current Radom factory literature indicates that the in-house German designation of this simplified model was P35/II(p). The serial numbers on parts were also repositioned, and although it is not visible, the quality of heat treatment gradually declined. A phosphate finish eventually replaced bluing on the VIS, and wooden ersatz grips replaced Bakelite, which had in turn replaced hard rubber; the wood grips coming at some time after the elimination of grip screw bushings. The trigger became a fabricated part, and the matted sighting ridge atop the slide was no longer milled off behind the rear sight. The magazine follower became a stamping. All of these changes were in order to simplify production, and none adversely affected the pistol's shooting qualities.

In December 1944 / January of 1945, Soviet forces were to occupy the city of Radom, but they found that the arms plant's inventory and equipment had been evacuated.

Wartime documents hint that the first choice for relocating the Radom plant was at FLUMO, an abbreviate term for Flugmotorenwerke-Ostmark. Flugmotorenwerke-Ostmark was an (underground) aircraft engine manufacturing facility located north of Steyr.

(Fig. 41-1) A memorial to the executed resistance workers as it stands today on factory grounds.
(Photograph courtesy of Waclaw Ustupski & Alicja Kopycka)

(Fig. 42-1) Visiting German officers inspect VIS production parts. (Photograph courtesy of Waclaw Ustupski)

The facility incorporated a satellite camp of the Mauthausen concentration camp. FLUMO was added to the Steyr umbrella of factories in 1943 due to inefficient productivity. Allied air raids forced the engine manufacturing to be first relocated to the Slovak town of Dubnica an later to the Skoda Works.

The lack of available space at the main Steyr facility and the references in the wartime documents to "Gusen" or "St. Georgen an der Gusen" indicates that Steyr manufactured parts and/or assembled VIS pistols at the Gusen concentration camp. This too was a satellite camp of Mauthausen concentration camp, divided among three labor camps, it was primarily utilized by Steyr. Gusen was known for munitions and production of small arms including the K-98 Mauser. The Gusen camps were a mere 40 kilometers from Steyr. The close proximity from Gusen to the Flugmotorenwerke-Ostmark facility also reinforces the location as a prime location to consolidate VIS production.

Increased Allied bombing made this option no longer viable by the summer of 1944. Factory documents indicate that evacuation was to a Steyr sub-plant which had been set up to produce automotive parts at the District Industrial Porcelain Factory in Znaim (Znojmo), Czechoslovakia, some 200 kilometers from Steyr. The works were known as Auto-Ersatzteilwerkstatte Znaim and retained this name while producing VIS pistols.

Production at this plant was to commence in November 1944, but production and assembly of "kits" was delayed into December. Plans were in the works to produce a yet more simplified version of the VIS with a formed sheet metal frame, to be known in-house as the P35/III(p).

Znaim production was plagued with problems from the onset; unskilled labor, labor shortages, material shortages and allied bombing made it impossible to achieve production quotas. A Steyr memo from December 29, 1944 instructs the factory to produce 78,000 pistol kits, by any means, utilizing on-hand parts. It is clear that Steyr opted to utilize all existing stockpiles before switching over production to the further simplified P35/III(p).

VIS production was to limp forward until nearly the end of the war at Steyr/Znaim. These last pistols were all made utilizing existing stockpiles of parts with traditionally manufactured frames and slides. These show further simplifications on a sporadic basis, such as solid rather than articulated captive recoil spring rods, an ersatz way of retaining the magazine catch on at least one pistol, and a change in markings. The P35/III(p) would never be produced.

P35/III(p) Technical Drawings

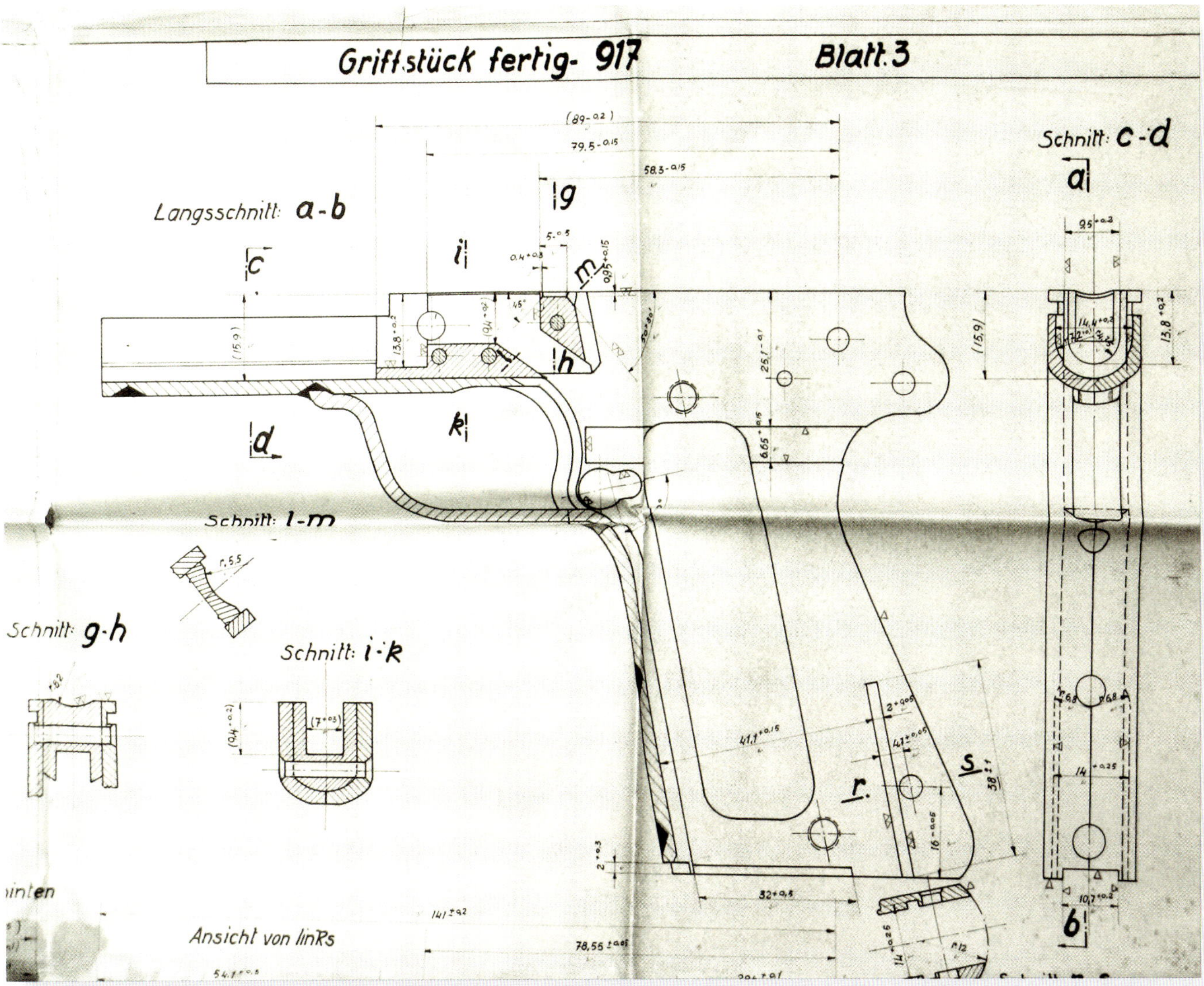

(Fig. 43-1) This drawing shows the frame to be of pressed and welded construction with machined camming surfaces (g-h and i-k) permanently pinned in. (Drawing courtesy of Waclaw Ustupski)

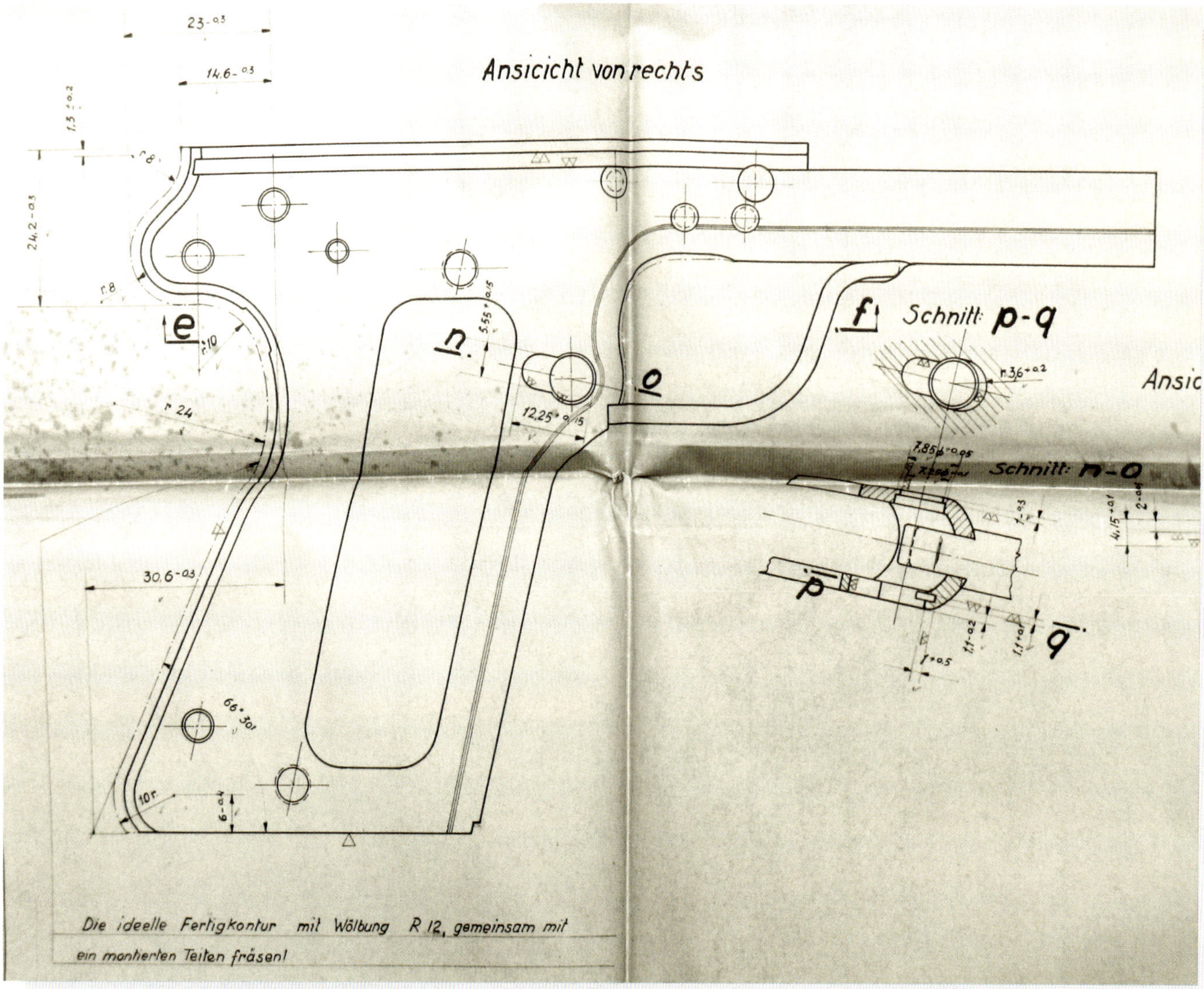

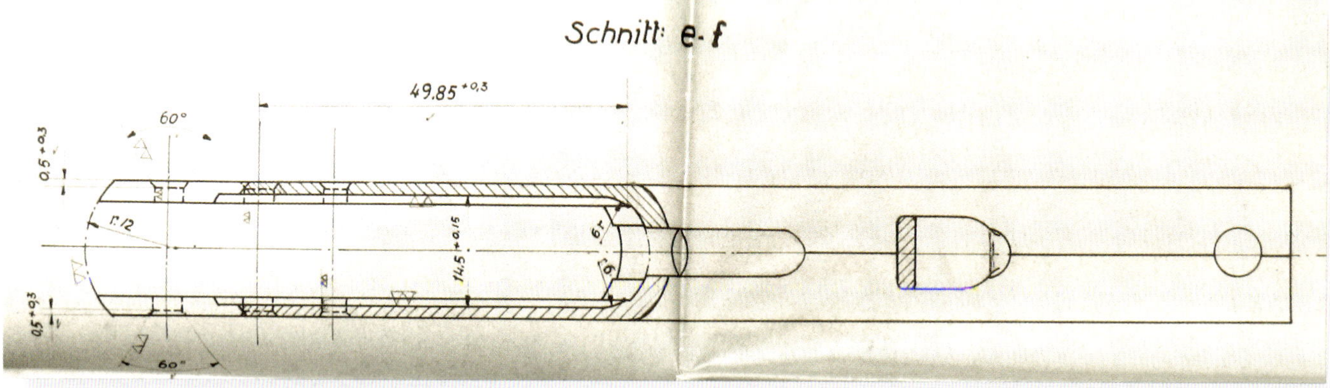

(Fig. 44-1) Views of the right side and a cutaway of the underside, of the projected stamped VIS frame. Drawings dated April 21, 1944. (Drawing courtesy of Waclaw Ustupski)

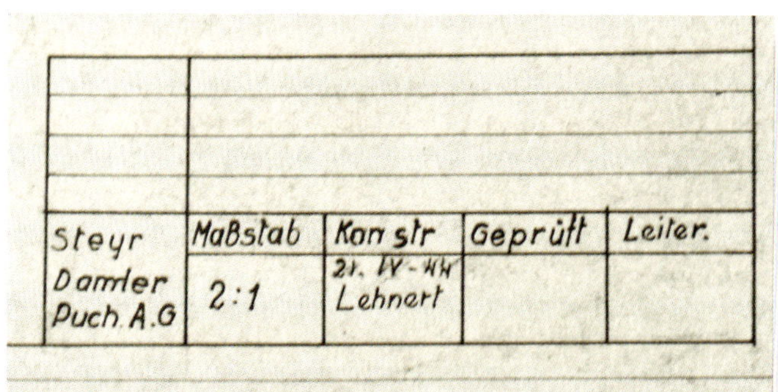

Select Wartime Factory Documents

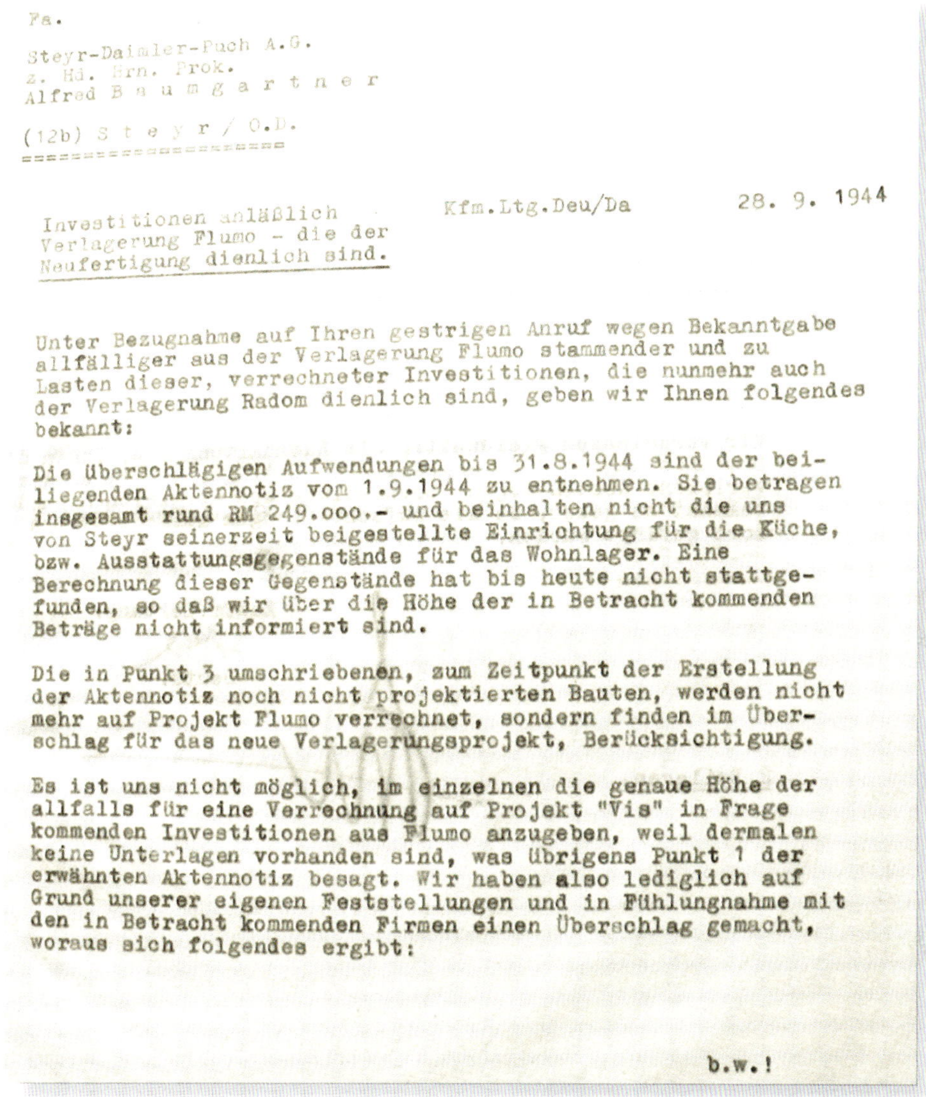

Fa.

Steyr-Daimler-Puch A.G.
z. Hd. Hrn. Prok.
Alfred B a u m g a r t n e r

(12b) S t e y r / O.D.
=====================

Investitionen anläßlich Verlagerung Flumo – die der Neufertigung dienlich sind.

Kfm.Ltg.Deu/Da

28. 9. 1944

Unter Bezugnahme auf Ihren gestrigen Anruf wegen Bekanntgabe allfälliger aus der Verlagerung Flumo stammender und zu Lasten dieser, verrechneter Investitionen, die nunmehr auch der Verlagerung Radom dienlich sind, geben wir Ihnen folgendes bekannt:

Die überschlägigen Aufwendungen bis 31.8.1944 sind der beiliegenden Aktennotiz vom 1.9.1944 zu entnehmen. Sie betragen insgesamt rund RM 249.000.- und beinhalten nicht die uns von Steyr seinerzeit beigestellte Einrichtung für die Küche, bzw. Ausstattungsgegenstände für das Wohnlager. Eine Berechnung dieser Gegenstände hat bis heute nicht stattgefunden, so daß wir über die Höhe der in Betracht kommenden Beträge nicht informiert sind.

Die in Punkt 3 umschriebenen, zum Zeitpunkt der Erstellung der Aktennotiz noch nicht projektierten Bauten, werden nicht mehr auf Projekt Flumo verrechnet, sondern finden im Überschlag für das neue Verlagerungsprojekt, Berücksichtigung.

Es ist uns nicht möglich, im einzelnen die genaue Höhe der allfalls für eine Verrechnung auf Projekt "Vis" in Frage kommenden Investitionen aus Flumo anzugeben, weil dermalen keine Unterlagen vorhanden sind, was übrigens Punkt 1 der erwähnten Aktennotiz besagt. Wir haben also lediglich auf Grund unserer eigenen Feststellungen und in Fühlungnahme mit den in Betracht kommenden Firmen einen Überschlag gemacht, woraus sich folgendes ergibt:

b.w.!

(Fig. 45-1) Cost analysis from Steyr dated September 28, 1944

English language synopsis: ...the facilities, that have not yet been established, are no longer calculated with the FLUMO project, but are part of the newly moved/relocated project. It is not possible for us, at this time, to calculate exactly the costs of the VIS project in relation to the FLUMO investments. We have only estimated costs on our calculations with the new firm.

Notes: This document hints to the fact the VIS production was originally to be paired with other manufacture at the FLUMO works. (Document courtesy of Waclaw Ustupski)

Bauarbeiten laut Beilage ausschließlich der Fundamente die für die Vis-Fertigung überhaupt nicht benötigt werden, ... RM 101.000.--

Elektro-Installationsarbeiten, ausschließlich der für die Einzelantriebe der Flumo-Maschinen bestimmten Stränge und Anschlußkästen, da für die neue Fertigung ausschließlich Maschinen mit Transmissionsantrieb Verwendung finden, rund .... RM 60.000.--

Installationsarbeiten im Wohnlager, Werksküche und Wasserinstallationen im Betrieb, sowie Preßluftleitungen .... RM 35.000.--

Verdunkelung des Wohnlagers, Küche, Büros, ausschließlich der noch zu erstellenden Verdunkelungsanlage für die Shedhallen, welch letzterer im Projekt "Vis" Berücksichtigung finden. .... RM 15.000.--

insgesamt RM 211.000.--

================================

Wir veranlassen gleichzeitig die Buchhaltung WKB, Ihnen die bis Ende August 1944 angefallenen Beträge, soweit es sich um verlorene Kosten wegen Übersiedlung und dergleichen handelt bekannt zu geben, da die Aufschreibungen ausschließlich dort geführt wurden.

Kaufm. Leitung:

Deutz.

2 Beilagen.

(Fig. 46-1) Cost analysis from Steyr dated September 28, 1944 (continued)

English language synopsis: Construction; except for the foundations (buildings) that are not necessary for the VIS manufacture. Electrical installations: Except for the FLUMO electrical machines connectors/wiring, considered that the new manufacture only requires machinery that operates on drive-shafts.

Notes: This rough costs estimation differentiates the VIS production from the other production, which was to be installed at FLUMO. An interpretation would be that buildings were not necessary as the VIS production was moved into hall 1, 2 and 3 of the existing Znaim factory. Electrical preparation was also not required as the VIS production machinery was shaft-driven and did not require electrical connections. (Document courtesy of Waclaw Ustupski)

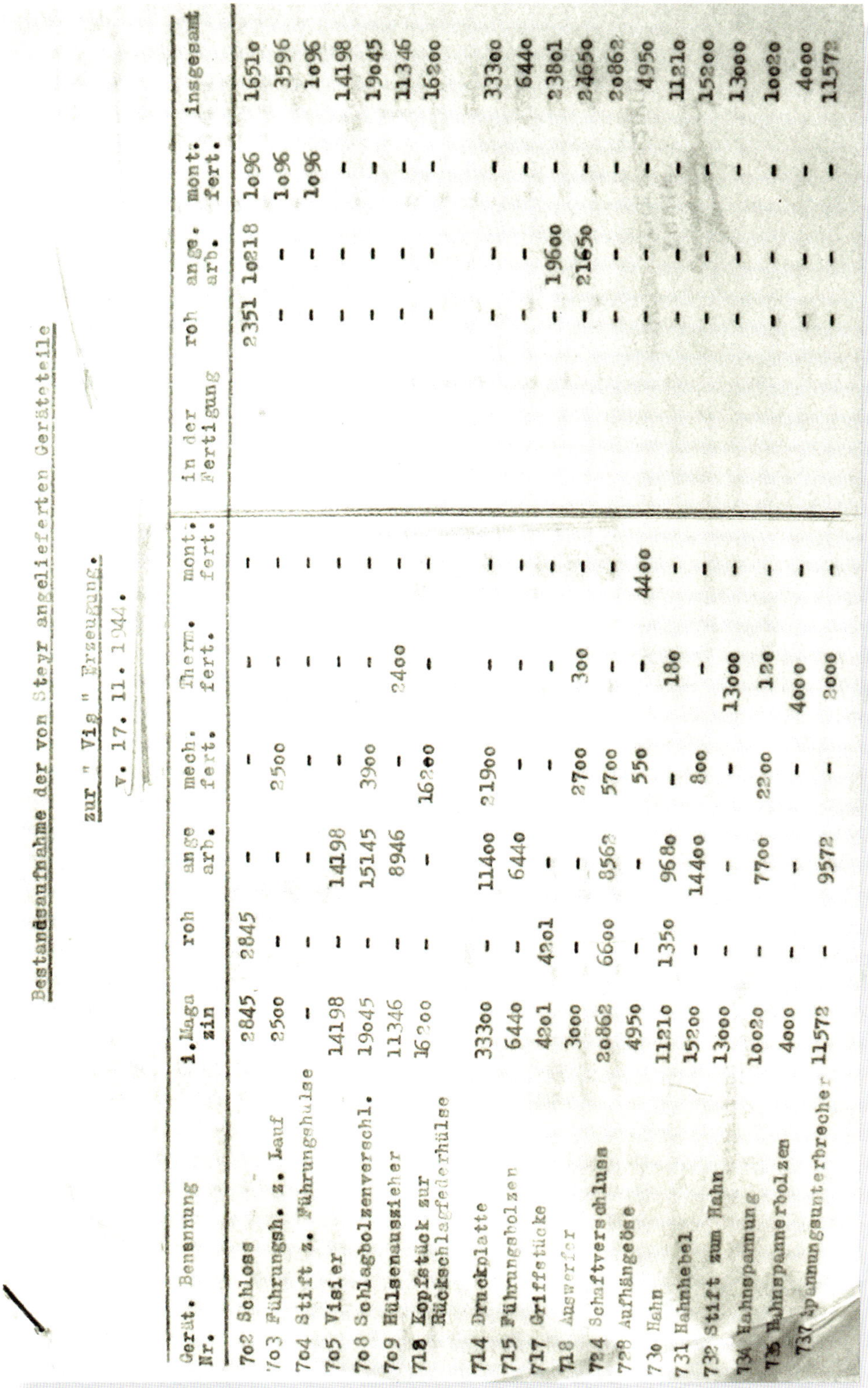

Bestandsaufnahme der von Steyr angelieferten Geräteteile

zur " Vis " Erzeugung.

v. 17. 11. 1944.

Gerät. Nr.	Benennung	1. Magazin	roh	ange arb.	mech. fert.	Therm. fert.	mont. fert.	in der Fertigung	roh	ange. arb.	mont. fert.	insgesamt
7o2	Schloss	2845	2845	-	-	-	-		2351	1o218	1o96	1651o
7o3	Führungsh. z. Lauf	25oo	-	-	25oo	-	-		-	-	1o96	3596
7o4	Stift z. Führungshulse	-	-	-	-	-	-		-	-	1o96	1o96
7o5	Visier	14198	-	14198	-	-	-		-	-	-	14198
7o8	Schlagbolzenverschl.	19o45	-	15145	39oo	-	-		-	-	-	19o45
7o9	Hülsenauszieher	11346	-	8946	-	24oo	-		-	-	-	11346
718	Kopfstück zur Rückschlagfederhülse	162oo	-	-	162oo	-	-		-	-	-	162oo
714	Druckplatte	333oo	-	114oo	219oo	-	-		-	-	-	333oo
715	Führungsbolzen	644o	-	644o	-	-	-		-	-	-	644o
717	Griffstücke	42o1	42o1	-	-	-	-		-	196oo	-	238o1
718	Auswerfer	3ooo	-	-	27oo	3oo	-		-	2165o	-	2465o
724	Schaftverschluss	2o862	66oo	8562	57oo	-	-		-	-	-	2o862
728	Aufhängeöse	495o	-	-	55o	-	44oo		-	-	-	495o
73o	Hahn	1121o	135o	968o	-	18o	-		-	-	-	1121o
731	Hahnhebel	152oo	-	144oo	8oo	-	-		-	-	-	152oo
732	Stift zum Hahn	13ooo	-	-	-	13ooo	-		-	-	-	13ooo
734	Hahnspannung	1oo2o	-	77oo	22oo	12o	-		-	-	-	1oo2o
735	Hahnspannerbolzen	4ooo	-	-	-	4ooo	-		-	-	-	4ooo
737	Spannungsunterbrecher	11572	-	9572	-	2ooo	-		-	-	-	11572

(Fig. 47-1) Inventory of parts supplied by Steyr – November 17, 1944 – for VIS production

Notes: This list was most likely drafted at Steyr and not Znaim as it lists parts in their various production stages both warehoused (Magazine) and in production (in der Fertigung). This reflects most likely the Steyr (Gusen) inventory and was relabeled as a shipping manifest. (Document courtesy of Waclaw Ustupski)

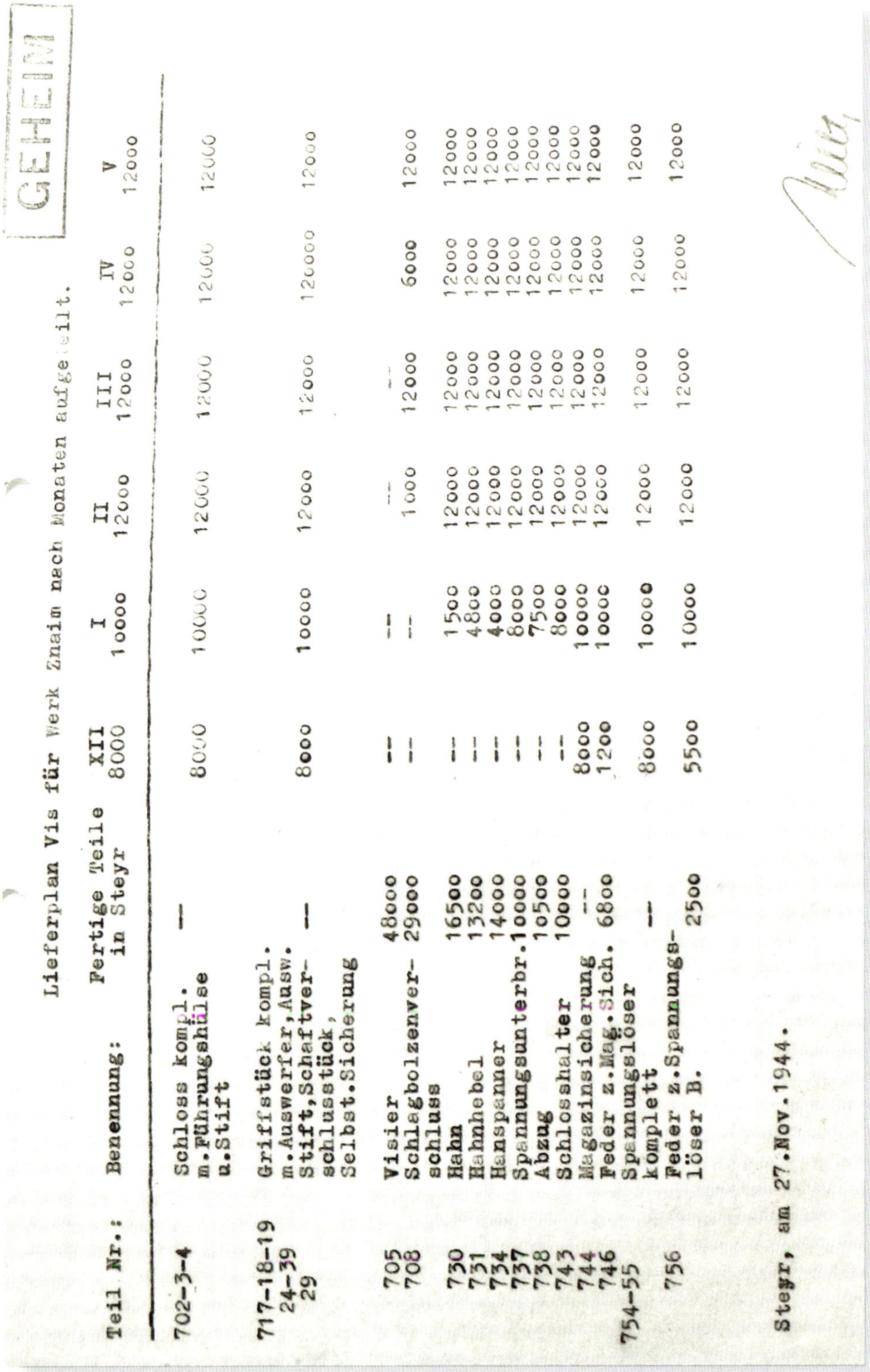

GEHEIM

Lieferplan Vis für Werk Znaim nach Monaten aufgeteilt.

Teil Nr.:	Benennung:	Fertige Teile in Steyr	XII 8000	I 10000	II 12000	III 12000	IV 12000	V 12000
702-3-4	Schloss kompl. m.Führungshülse u.Stift	—	8000	10000	12000	12000	12000	12000
717-18-19 24-39 29	Griffstück kompl. m.Auswerfer,Ausw. Stift,Schaftverschlusstück, Selbst.Sicherung	—	8000	10000	12000	12000	12000	12000
705	Visier	48000	--	--	--	--		
708	Schlagbolzenverschluss	29000	--	--	1000	12000	6000	12000
730	Hahn	16500	--	1500	12000	12000	12000	12000
731	Hahnhebel	13200	--	4800	12000	12000	12000	12000
734	Hanspanner	14000	--	4000	12000	12000	12000	12000
737	Spannungsunterbr.	10000	--	8000	12000	12000	12000	12000
738	Abzug	10500	--	7500	12000	12000	12000	12000
743	Schlosshalter	10000	--	8000	12000	12000	12000	12000
744	Magazinsicherung	--	8000	10000	12000	12000	12000	12000
746	Feder z.Mag.Sich.	6800	1200	10000	12000	12000	12000	12000
754-55	Spannungslöser komplett	—	8000	10000	12000	12000	12000	12000
756	Feder z.Spannungslöser B.	2500	5500	10000	12000	12000	12000	12000

Steyr, am 27.Nov.1944.

(Fig. 48-1) Delivery plan/schedule for the Znaim works, divided into months – set up by Steyr on November 27, 1944

Notes: The Roman numerals indicate the months starting with XII for December 1944, followed by I for January 1945, etc. Although documents indicate that Znaim did assemble pistols, there is no indication that any barrels or magazines were manufactured or on hand at Znaim. Final assembly with barrels and magazines continued at Steyr (Gusen) just as it had during the Radom production.

Of importance are also the "on hand" Steyr parts that are not initially manufactured at Znaim like the rear sight (Visier), hammer (Hahn), etc. These earlier manufactured parts were finished at Steyr and are found on late pistols in blue finish, while the Znaim finished parts (frame, slide, etc.) are in a phosphate finish. Aside from originating from another source, there was a practical reason to have those parts blued as a phosphate finish added to the dimensions of the parts and possibly hindered operation. (Document courtesy of Waclaw Ustupski)

Znaim, am 1.12.44

GEHEIM

FERTIGUNGSPLAN für das Werk Znaim /mech.Fertigung/ nach Monaten aufgeteilt
mit 2 Wochen Vorlauf.

Teil-Nr.:	Benennung!	Fert.Teile i.Steyr	Fert.Teile i.Znaim	Summe:	XI. -	XII. 8000	I. 10000	II. 12000	III. 12000	IV. 12000	V. 12000
702	Schloss	-	1096	1096							
703	Führungshülse	-	3596	3596	3000	9000	11000	12000	12000	12000	12000
704	Stift z.Führungshülse	30000	1096	31096	500	9000	11000	12000	12000	12000	12000
705	Visier	48000	—	48000	-	—	—	4000	12000	12000	12000
708	Schlagbolzenverschl.	29000	3900	32900	—	—	—	—	—	12000	12000
717	Griffstück	—	457	457	—	—	—	2000	12000	12000	12000
718	Auswerfer	—	3000	3000	3600	9000	11000	12000	12000	12000	12000
719	Auswerferstift	18000	—	18000	1000	9000	11000	12000	12000	12000	12000
724	Schaftverschlussstück	7000	5700	12700	—	—	6000	12000	12000	12000	12000
729	Stift z.Schaftverschlst.	7800	—	7800	—	—	11000	12000	12000	12000	12000
730	Hahn	16500	—	16500	—	5000	11000	12000	12000	12000	12000
731	Hahnhebel	13200	800	14000	—	—	7500	12000	12000	12000	12000
734	Hahnspanner	14000	2300	16300	—	—	10000	12000	12000	12000	12000
737	Spannungsunterbrecher	10000	2000	12000	—	—	8000	12000	12000	12000	12000
738	Abzug	10500	500	11000	—	1000	11000	12000	12000	12000	12000
739	Selbst.Sicherung	—	7300	7300	—	2000	11000	12000	12000	12000	12000
743	Schlosshalter	10000	1000	11000	—	7000	11000	12000	12000	12000	12000
744	Magazinssicherung	—	600	600	—	2000	11000	12000	12000	12000	12000
746	Feder z. - " -	6800	—	6800	3500	9000	11000	12000	12000	12000	12000
754	Spannungslöserbolzen	1300	12500	13800	—	6000	11000	12000	12000	12000	12000
755	Flügel z.Spannungslöser	--	9400		—	—	10000	12000	12000	12000	12000
754-55	Spannungslöser kompl.	--	7400	7400	—	6000	10000	12000	12000	12000	12000
756	Feder z.Spannungslöser	2500	—	2500	1500	9000	11000	12000	12000	12000	12000
758	Bügel	x10500	500	11000	—	2000	11000	12000	12000	12000	12000
759	Druckstück	x10500	500	11000	—	2000	11000	12000	12000	12000	12000
761	Niete	21000	1000	22000	—	4000	22000	12000	12000	12000	12000
702-3-4	Schloss kompl.	--	1096	1096	1000	8500	10500	12000	12000	12000	12000
702-717	Schloss m.Griffstück	--	x259	259	1700	8500	10500	12000	12000	12000	12000
717-18-19	Griffstück m.Ausw.	--	x457	457	1500	8500	10500	12000	12000	12000	12000
717-24-29-39	" m.Verschl. Stück u.Sicherung	--	x457	457	1500	8500	10500	12000	12000	12000	12000

Steyr, 27.Nov.1944.

(Fig. 49-1) Production schedule for the Znaim works, divided into months – issued at Znaim on December 1, 1944.

Notes: The Roman numerals indicate the months starting with XI for November 1944, followed by XII for December 1944, etc. The "Fert.Teile" columns indicate completed parts on hand from Steyr or Znaim. These indicate only where the parts came from and included without doubt predominately Radom manufactured parts, note the lack of a barrel (Lauf) column.

Note that only 457 finished frames are on hand. Note also the overly ambitious production numbers for the frame. It is unclear if, at this date, these are still to be traditional forged frames or the newer sheet metal frame. (Document courtesy of Waclaw Ustupski)

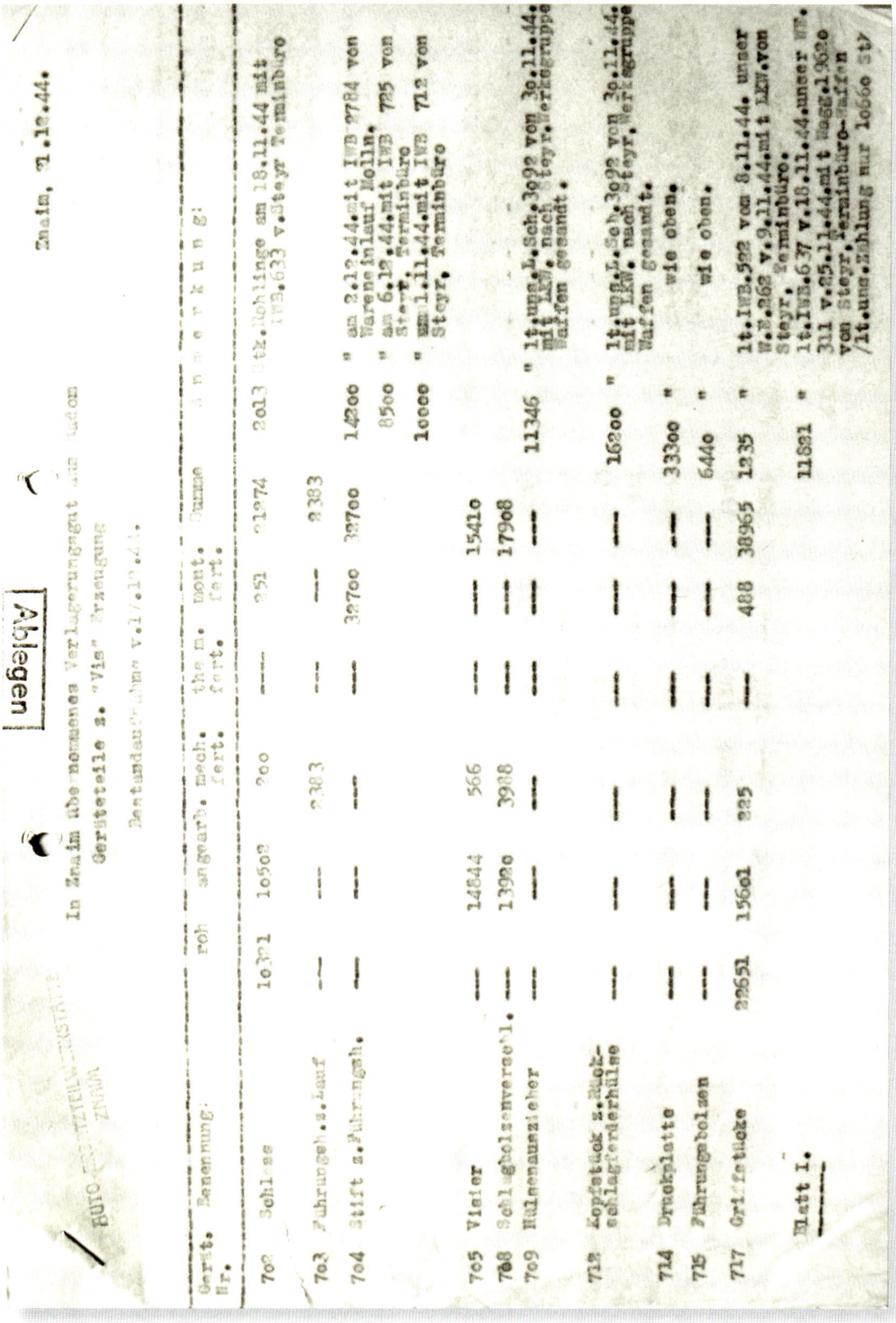

Ablegen

AUTO-ERSATZTEILWERKSTÄTTE ZNAIM

Znaim, 21.12.44.

In Znaim übernommenes Verlagerungsgut aus Radom
Gerätteile z. "Vis" Erzeugung

Bestandaufnahme v.17.12.44.

Gerät. Nr.	Benennung:	roh	angearb.	mech. fert.	therm. fert.	mont. fert.	Summe	A n m e r k u n g:
702	Schlosse	10321	10502	200	—	251	21274	2013 Stk.Rohlinge am 18.11.44 mit IWB.633 v.Steyr Terminbüro
703	Führungsh.z.Lauf	—	—	2383	—	—	2383	
704	Stift z.Führungsh.	—	—	—	—	32700	32700	14200 " am 2.12.44.mit IWB 2784 von Wareneinlauf Molln. 8500 " am 6.12.44.mit IWB 725 von Steyr, Terminbüro 10000 " am 1.11.44.mit IWB 712 von Steyr, Terminbüro
705	Visier	—	14844	566	—	—	15410	
708	Schlagbolzenverschl.	—	13920	3988	—	—	17908	
709	Hülsenauszieher	—	—	—	—	—	—	11346 " lt.ups.L.Sch.3092 von 30.11.44. mit LKW. nach Steyr.Werksgruppe Waffen gesandt.
712	Kopfstück z.Rückschlagförderhülse	—	—	—	—	—	—	16200 " lt.ups.L.Sch.3092 von 30.11.44. mit LKW. nach Steyr,Werksgruppe Waffen gesandt.
714	Druckplatte	—	—	—	—	—	—	33300 " wie oben.
715	Führungsbolzen	—	—	—	—	—	—	6440 " wie oben.
717	Griffstücke	22651	15601	225	—	488	38965	1235 " lt.IWB.522 vom 8.11.44. unser W.E.262 v.9.11.44.mit LKW.von Steyr, Terminbüro. 11821 " lt.IWB.637 v.18.11.44.unser WE. 311 v.25.11.44.mit Wagg.19620 von Steyr, Terminbüro-Waffen /lt.uns.Zählung nur 10660 St/

Blatt I.

(Fig. 50-1) Stocks from Radom, accepted at Znaim – for production of VIS – December 17, 1944

Notes: The parts are inventoried by their stage of completion:
Roh = raw material stage, first manufacturing stage
angearb. = in production
mech.fert. = machining completed
therm.fert. = heat treating completed
mont.fert. = ready for assembly

(Document courtesy of Waclaw Ustupski)

AUTO-ERSATZTEILWERKSTÄTTE
ZNAIM

Znaim, den 22. XII. 1944

Aktennotiz

über den Besuch in der Auto-Ersatzteilwerkstätte Znaim
am 22. XII. 1944.

Anwesende:

Hr. Dir. Janku
Hr. Prok. Schweiger
Hr. Deutz
Hr. Ing. Mayer
Hr. Jochmann

1./ Fertigungs- und Liefersituation:

Die Ausbringung des gesamten Gerätes wird hauptsächlich diktiert von der Lieferung zusammengepaßter Schloß- und Griffstückteile. Von diesem Aggregat hat Znaim bis inkl. 22.12. 850 Satz fertiggestellt. Von diesen 850 Sätzen sind

450 bereits in Molln
2oo Satz abgegangen am 22.12 Vormittag mit LKW über Wien und gehen weitere
2oo Satz am 22.12. Nachmittag mit LKW direkt nach Steyr.

Es ist angeordnet, daß bei allen Engpaßstellen und hiezu gehört hauptsächlich die Zusammenpaßarbeit, die Feiertage über voll gearbeitet wird, wobei der 24.XII. als Einschichttag zu halten ist.

Bei der derzeitigen Tagesausbringung von rund 1oo Satz je Tag, ergibt sich bis 26.12. früh eine Fertigstellung von weiteren 3oo Satz. Herr Deutz wird veranlassen, daß diese 3oo Satz noch am 26. mit LKW nach Steyr abgehen. Die vorläufige weitere Tageslieferung von 1oo Stück muß ab 27.12. als Expreßgutsendung nach Steyr abgerichtet werden. Diese Expreßsendungen sind nach Angabe des Herrn Deutz von Seiten des Rüstungskommandos Mödling und der DR. trotz der bis 1.I. dauernden Sperre, zugesagt worden.

Bei allen übrigen in Znaim zu fertigenden Teilen, besteht unter Berücksichtigung der Minderausbringung von Griffstück und Schloß keine unmittelbare Programmgefährdung.

Es wurden nochmals sämtliche in Znaim zu fertigenden Teile eingehendst durchbesprochen und dem Programm entsprechend die noch vorzunehmenden Einrichtungsarbeiten wie folgt festgelegt:

- 2 -

(Fig. 51-1) Memo on the visit to the Auto-Ersatzteilwerkstatte Znaim on December 22, 1944.
Manufacturing and delivery situation I

English language synopsis:
The production of the kit consists of the assembled slide and frame components. Up to December 22, Znaim has delivered 850 of such kits. 450 of these are already in Molln, 200 have left by truck on a route through Vienna and 200 kits will ship by truck directly to Steyr. With daily production reaching 100 kits, there will be 300 kits available on December 26. Mr. Deutz will arrange that these will ship that day by truck to Steyr. From the 27th and on, daily production of 100 kits will have to be shipped as express freight to Steyr.

Notes: It is obvious from the urgency listed that Steyr was in (desperate) need of kits for final assembly. Molln is a small town about 30 km from Steyr. It is unknown why the kits were shipped there, this may have been an assembly location. We can speculate that various routes were taken in order to avoid large shipments from being destroyed by Allied aircraft. (Document courtesy of Waclaw Ustupski)

(Fig. 52-1) Memo on the visit to the Auto-Ersatzteilwerkstatte Znaim on December 22, 1944. Manufacturing and delivery situation (II continued)

Für Teil	705	März 1945
	708	Feber 45
	730) 731) 734) 738) 746)	10. Jänner 45
	754	15. Jänner 45
	755	Feber 45
	758) 759)	10. Jänner 45
	761	Feber 45

Für das Teil 755 Spannungslöser-Flügel fehlt in Znaim die kompl. Einrichtung. Dieser Teil wurde seinerzeit in Warschau gefertigt und müssen die Vorrichtungen in Steyr gesucht werden. Falls diese nicht auffindbar sind, muß die Anfertigung in Steyr aufgegeben werden, und zwar so, daß sie Znaim bis spätestens 15. Feber zur Verfügung stehen.

Für Teil 758 Abzugbügel fehlen ebenfalls die Einrichtungen. Davon fertigt Znaim bis 10.I. das Biegewerkzeug und das Abschneidwerkzeug. Das Tiefziehwerkzeug (das vermutlich noch auf einer Maschine aufgespannt ist) muß in Steyr gesucht werden. Falls nicht vorhanden, muß es ebenfalls in Steyr bestellt werden und Znaim bis spätestens 10.I. zur Verfügung stehen.

Sämtliche Betriebsmittelzeichnungen nach neuem Operations-Plan für Teil 758 sind Znaim in 2-facher Ausfertigung zu senden.

2./ <u>Begründung der Nichtprogramm-Einhaltung.</u>

Zur Zeit meines letzten Besuches in Znaim mit 9.12.44 wurde von Hrn. Ing. Mayer eine Tagesausbringung von 1oo - 12o Stück zusammengepaßten Griffstück- und Schloß-Sätzen zugesagt. Diese Zusage ist bei weitem nicht eingehalten worden, und wird von Hrn. Ing. Mayer wie folgt begründet:

Die in der zweiten Dezember-Woche erreichte Tagesausbringung von 1oo Stück war, um das vorgeschriebene Programm halten zu können, ungenügend. Die Steigerung auf eine höhere Tages-Stückzahl sollte durch Neuansetzen von Schlossern aus der Gruppe Hintermayer (Transmissionsaufbau) möglich gemacht werden. Es sind zur Zeit insgesamt 13 Schlosser oder angelernte Schlosser in der Montage-Gruppe tätig, die das Ausfertigen der Griffstücke sowie das Zusammenpassen mit dem kompletten Schloß durchzuführen haben.

Die Montagearbeiten wurden mit den Radomer Rest-Teilen begonnen, wobei sich vor allem bei den neuangesetzten Leuten Arbeitsschwierigkeiten infolge Unkenntnis der

./.

English language synopsis: The entire manufacturing line for part 755, decocking lever, is missing at Znaim. This part used to be made in Warsaw and the installations should be sought at Steyr. In case the machinery can not be located, the parts should be produced at Steyr so that these are available by no later than February 15. The manufacturing line for part 758 is also missing…

2) Failure to follow delivery schedule

At the time of my visit to Znaim on September 12, '44, the projections of Mr. Mayer were 100 to 120 kits daily. This projection has for long not been obtained. In the second week of December assembly reached 100 kits daily, an insufficient amount to uphold the delivery schedule. Daily production should be reached with the addition of slide manufacturers (machinists) from the group Hintermayer. There are now a total of 13 machinists and machinists in training working in assembly of frame and slide (fitting the parts together). Assembly of the other parts from Radom has started and has caused problems with the newly appointed workers.

Notes: Note that the writer refers to the part being made in Warsaw and not Radom. The Germans used the old Polish parts numbers, yet it appears that the numbers were slightly updated once manufacturing simplifications were instilled. Ultimately it is not clear where the decocking lever ended up being manufactured. Late war phosphate finish guns have either blue decockers or phosphate decockers. The blue finish parts on late phosphate guns are thus early parts supplied by Steyr.

The September date listed should not be interpreted as an indicator that production was in place at that time. It was most likely a projected estimate. Note that machinists were working at fitting. Fitting in this case is not final filing, but rather machining the parts so that they are compatible; this may be an indicator that not all parts were easily interchangeable. (Document courtesy of Waclaw Ustupski)

\- 3 -

Vorteile ergaben. Es ist klar, daß bei diesen Montagearbeiten die Leute viel länger zum Anlernen benötigen, als ein normaler Ausfertigungs-Schlosser. Die Steigerung der Tagesausbringung über 100 Stück hinaus wäre auch möglich gewesen, wenn die Griffstück-Fertigung selbst für die Montage einwandfreie Teile gebracht hätte.

Es muß hier festgehalten werden, daß die gesamte Vis-Fertigung, also Einsteller, Maschinenarbeiter und Schlosser, mit Leuten besetzt ist die mit der Fertigung bisher, geschweige denn mit der Funktion der Pistole vertraut waren.

Durch den Ausfallé des für Znaim vorgesehenen, verunglückten Kontroll-Leiters Antensteiner, sind die Einrichtearbeiten in der Kontrolle sowie das Aufziehen der notwendigen Zwischen-Kontrollstellen verzögert worden und konnte erst nach Eintreffen des Behelfsmeisters Dietrich aus Gusen, der sich ebenfalls mit der Materie vertraut machen mußte, nach und nach aufgezogen werden. Vertrauend auf die Fähigkeiten der Einsteller, mußte man sich mit Rücksicht auf das gestellte Programm dazu entschließen, die Teile ohne 100 % Kontrolle durchzuziehen.

Obwohl die ~~Kontrolle~~ Montage 100 und auch mehr Stücke hätte machen können, hat der Ausfall aus der Erzeugung, infolge nicht entsprechender Maßhältigkeit dies unmöglich gemacht. Es haben sich verschiedentlich sehr schwerwiegende und störende Nacharbeiten ergeben, die ja größten Teils von Hand aus durchgeführt werden mußten und daherdas katastrophale Absinken der Montage-Ausbringung zur Folge hatten.

Zum Arbeitseinsatz selbst, der beim letzten Besuch des Herrn Dir. Janku von diesem bezüglich der Nachtschicht-Besetzung beanstandet wurde ist zu bemerken, daß sich bis zum heutigen Tag gegenüber dem damaligen Zustand, nichts wesentliches geändert hat. Der Fertigung konnten die für die Programm-Einhaltung notwendigen Facharbeiter, Maschinenarbeiter und Maschinenarbeiterinnen nicht zur Verfügung gestellt werden. Über Major Glöckl wurden in der vorigen Woche dem Betrieb 16 männliche, 4 weibliche und 2 jugendliche, polnische Arbeitskräfte zugeteilt. Es muß festgehalten werden, daß sich darunter keine Industriearbeiter, geschweige denn Facharbeiter befunden haben, sondern daß es sich lediglich um Kräfte aus der Landwirtschaft handelt. Das Anlernen dieser wenigen Menschen bringt unter den gegebenen Verhältnissen, eher Betriebsstörung mit sich, als die erforderliche Entlastung.

(Fig. 53-1) Memo on the visit to the Auto-Ersatzteilwerkstatte Znaim on December 22, 1944. Manufacturing and delivery situation (III continued)

English language synopsis:
It is obvious that the (Znaim) workers need more training than the usual factory workers. Production increases above 100 kits per day would have also been possible if the supplied frames were faultless.
It should be noted that the entire work force (machinist, adjusters, etc.) was until recently unfamiliar with manufacturing, let alone the functions and manufacturing of a pistol.

The loss (due to an accident) of the appointed quality control leader has caused delays in final quality control as well as setting up intermediate QC checks. These will only be established after the arrival of Dietrich from Gusen, he too will have to get acquainted with the manufacturing line. Relying on the competence of the adjusters, and following the production schedule, it was necessary to pass the kits without 100 percent quality control. The incompatible tolerances of the parts have made it impossible to follow production guidelines of 100 or more kits. There has been a disturbing amount of post production work done, much by hand and in various ways, which all lead to a catastrophic decline in assembly numbers.

Regarding the labor force. There have been no changes since Mr Janku's visit whereas he complained about the night shift. The necessary production workers (machinist, fitters, adjusters, etc.) have not been made available according to the plan. Last week 16 Polish men, four Polish women and two Polish youngsters were assigned through Major Glockl. None of these were factory workers but came from an agricultural background. Training the few workers brings interruptions to the production process instead of alleviating the process.

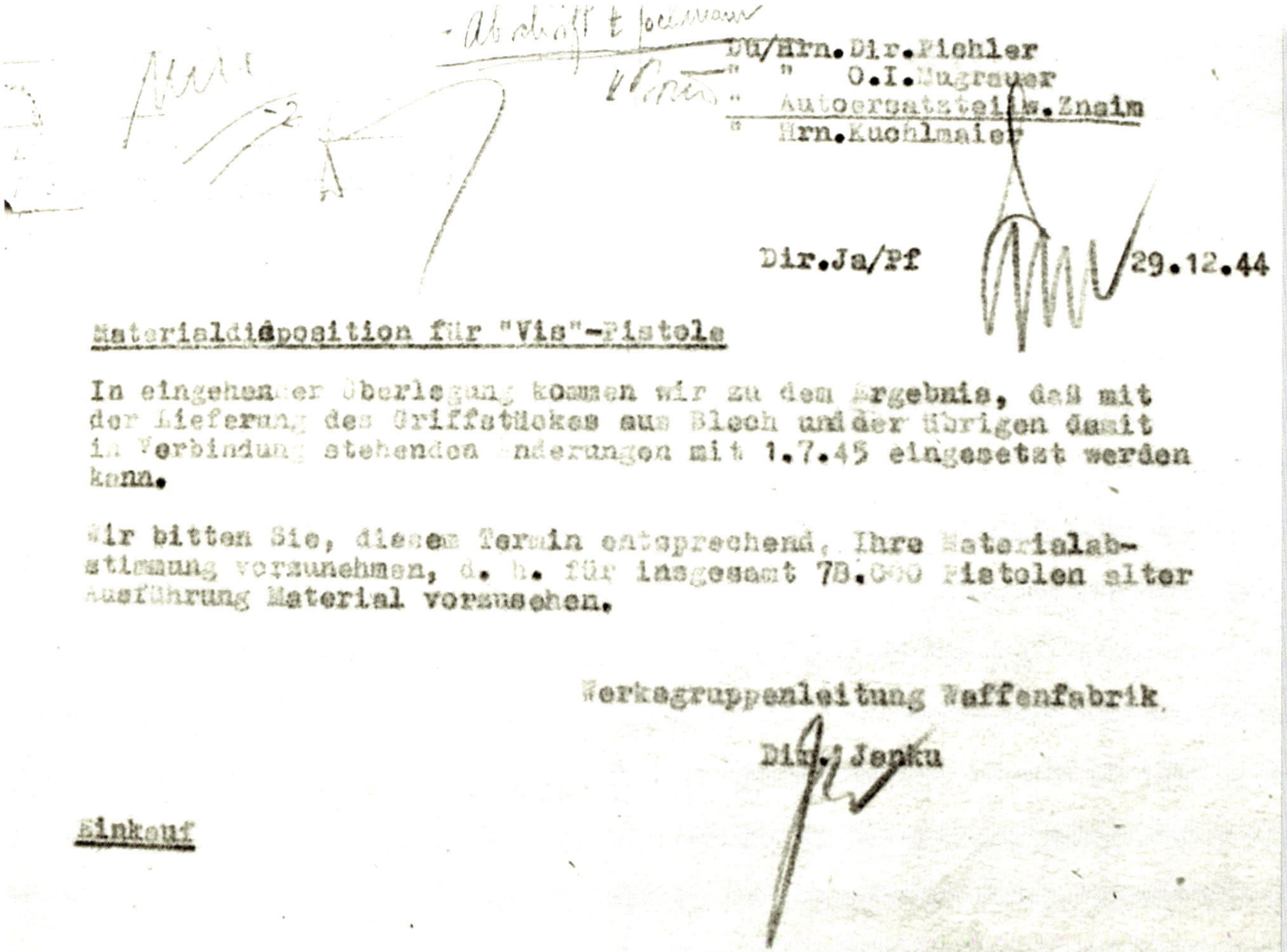

DU/Hrn. Dir. Pichler
" " O.I. Hugrauer
" Autoersatzteile W. Znaim
" Hrn. Kuchlmaier

Dir. Ja/Pf 29.12.44

Materialdisposition für "Vis"-Pistole

In eingehender Überlegung kommen wir zu dem Ergebnis, daß mit der Lieferung des Griffstückes aus Blech und der übrigen damit in Verbindung stehenden Änderungen mit 1.7.45 eingesetzt werden kann.

Wir bitten Sie, diesem Termin entsprechend, Ihre Materialabstimmung vorzunehmen, d. h. für insgesamt 78.000 Pistolen alter Ausführung Material vorzusehen.

Werksgruppenleitung Waffenfabrik

Dir. Jenku

Einkauf

(Fig. 54-1) Memo dated December 29, 1944

English language synopsis: We have concluded that we can start with deliveries of the sheet-metal frames and related changes on January 7, 1945. We request that you liquidate your materials, in other words, that you provide 78,000 pistols from old existing parts.

Notes: The 78,000 pistol production number is unrealistically ambitious, seeing the problems in getting assembly and production streamlined. (Document courtesy of Waclaw Ustupski)

Znaim, den 3. Jänner 1945.

Herren
Meister Karall und Perkounigg,
Halle I.

Betrifft: Fertigungsprogramm für Jänner 1945.

Sie haben laut Programm in diesem Monat von nachstehenden Teilen folgende Gutstückzahlen zu fertigen:

			Tagessoll /ohne Sonntage/
70[illegible]	Schloss	14.528	560 Stück
703	Führungshülse	14.404	560 "
718	Auswerfer	10.900	420 "
724	Schaftverschlusstück	5.300	210 "
729	Stift z. Schaftverschlusstück	10.200	400 "
730	Hahn	1.500	60 "
731	Hahnhebel	4.000	160 "
734	Hahnspanner	2.700	110 "
737	Spannungsunterbrecher	6.000	230 "
738	Abzug komplett	7.000	270 "
739	Selbsttätige Sicherung	10.700	420 "
743	Schlosshalter	6.400	250 "
744	Magazinsicherung	14.540	560 "
746	Feder zur Magazinsicherung	11.200	440 "
754	Spannungslöserbolzen	4.200	165 "
754/55	montiert	10.600	410 "
758	Bügel	7.000	270 "
759	Druckstück	7.000	270 "
761	Niete	14.000	540

Sie haben mit den, Ihnen zur Verfügung stehenden Mitteln, alles daran zu setzen, diese Stückzahlen zu erreichen.
Von Ihrer Seite aus muss die Einteilung in der Werkstätte so getroffen werden, dass der Ausstoss auch tatsächlich erreicht werden kann.

Für die bekannten Engpassteilen, die als schrittmachend der Fertigung voran gehen, erhalten Sie in den nächsten Tagen die erforderliche Anzahl von Einstellern und Schlossern, soweit sie in Steyr frei gemacht werden konnten.
Es ist vor allem besonders darauf zu achten, dass die Fertigung nach dem Taktsystem abläuft, das heisst bei Vermeidung auch jeder kleinsten Anhäufung, bei der Maschine müssen die Teile von Hand zu Hand weitergegeben werden.
Nachdem in kürzester Zeit weitere Maschinenarbeiter zur Verfügung gestellt werden, müssen Sie jetzt schon überlegen, wo und welche Leute gebraucht werden, damit der Arbeitseinsatz laut Programm eingehalten werden kann.

Durchschrift an

Herrn Prokurist Schweiger,
Herrn Brunner, Arbeitsvorbereitung.

Ing. Mayer.

(Fig. 55-1) Znaim internal directorate issued to the production overseers of Hall 1 – dated January 3, 1945

English language synopsis: You have to produce this month the following parts as per the production schedule. (list of parts + quantities) You have to do everything possible, with your available resources, to achieve these production numbers. From your part, the organization in the production plant has to be so that these numbers can be realistically achieved. For the known production problems, you will receive within days the required calibrators and machinists as far as they can be made available from Steyr.
It is especially important to pay attention that production follows the proper course; to avoid even the smallest accumulation of parts near the (production) machines, parts will have to passed on hand to hand. Additional machinists will be made available in the near future but you are already to determine where and what type of additional workers are required in order to maintain the production schedule.

Fa.
Steyr-Daimler-Puch A.G.
z. Hd.Hrn. Dir. Janku

(12b) S T E Y R / O.D.

Werk-Steyr

Prok.Schw./Da 5. I. 1945

Entsprechend der Aktennotiz vom 22.12.44 ist es notwendig schon heute die großen Schwierigkeiten aufzuzeigen, die im Anlauf der Fertigung bestehen und aller Voraussicht nach die Einhaltung des Jänner-Programms unmöglich machen werden.

1./ <u>Arbeitseinsatzlage:</u>

Die Arbeitseinsatzlage muß als äußerst schlecht bezeichnet werden. Es ist bereits wieder eine Woche des Monats Jänner verstrichen, ohne daß die von Steyr schon vor Neujahr zugesagten 200 Arbeitskräfte hier eingelangt sind. Es muß damit gerechnet werden, daß falls sie Anfang nächster Woche eintreffen noch immer bis zur Aufnahme, Einweisung, Arbeitsunterweisung, also bis zu einer halbwegs brauchbaren Leistung eine weitere Woche verstreicht, so daß diese 2oo Arbeitskräfte im Monat Jänner höchstens 14 Tage (2.Hälfte) als im Einsatz befindlich gerechnet werden können.

Noch viel schlimmer als das Vorhergesagte ist die Situation bezüglich der Einsteller. Wie Ihnen bereits mitgeteilt, haben wir zwischen Weihnachten und Neujahr nicht weniger als 4 Einsteller verloren, und zwar:

Riegler u. Niedermayr durch Abzug von Seiten Flumo
Sailer u. Stöcklmayr infolge Krankheit.

Außerdem ist Meister Perkounig ebenfalls infolge Krankheit ausgefallen. Der Abzug der Einsteller Riegler und Niedermayr hat in der Halle 1 eine katastrophale Lage hervorgerufen, weil

1./ ein bedeutendes Absinken in der Schloßfertigung eingetreten ist und

2./ das Aufziehen der Kleinteile-Fertigung derzeit überhaupt nicht möglich ist.

Aus den beigeschlossenen 3 Beilagen des Hrn. Ing. Mayer ersehen Sie die tatsächliche Einstellersituation in der

./.

(Fig. 56-1) Labor report dated January 5, 1945.
English language synopsis: As previously mentioned in a memo, he wants to make it clear that there are large problems with the production which will make it impossible to reach the January production target.

Situation of the labor force: This is very bad. Yet another week has passed without the arrival of the 200 workers appointed by Steyr. Even if they arrive early next week, it will be another week lost in order to assign and train them. Those 200 workers will be at best capable of working two weeks in January.

Worse is the situation with the machine calibrators, four were lost (two sick and two departed / were transferred). Slide manufacture has consequently diminished and starting production on some of the smaller parts is now impossible. (Document courtesy of Waclaw Ustupski)

Du/ Hrn. Dir. Pichler
Hrn. Prok. Schweiger

Werksgruppe Waffen

Betriebsleitung Znaim

Dir. J/E

19.1.1945

Ausbringung.

Der von Ihnen hierher entsandte Herr B r u n n e r legt mir das bisherige Lieferprogramm (Jänner) vor und muß ich feststellen, daß in den Tagen seit meinem letzten Besuch am 12./13. in der Ausbringung der wichtigsten Teile, wie Schloß und Griffstück, neuerdings keinerlei Fortschritt zu verzeichnen ist. Ich habe bei meinem letzten Besuch doch ausdrücklichst angeordnet, daß mir jede Betriebsstörung, die sich wesentlich auf die Ausbringung auswirkt, sofort zu melden ist. Ihre bisherigen Meldungen, speziell die vom heutigen Tage bezüglich des Maschinenausfalles beim Griffstück können nach Ihren eigenen Angaben nicht als Ursache der Minderlieferung angesehen werden.

Beim Schloß wurde während meines Besuches eine Ausbringung von 120 - 140 Stück je Tag angegeben und auch von Ihnen ab 10. bis 12. bis zu 150 Stück geliefert. Ab 13. erfolgt nun ein Abfall auf 100, bis 16. auf 47 Stück, am 17. Null-Lieferung und ist außerdem festzustellen, daß Sonntag, den 14. nicht gearbeitet wurde. Der seinerzeit während meiner Anwesenheit aufgenommene Umlauf war doch ausreichend, um die angegebene Lieferstückzahl je Tag bis zur endgültigen Einrichtung des Revolverzwischenstranges aufrecht zu erhalten. Es ist daher Ihr Ausbringungsausfall unerklärlich und muß ich Sie um ausführlichen Bericht bitten.

Beim Griffstück ist wohl Sonntag gearbeitet worden und im Durchschnitt die in der Notiz vom 13.1. festgelegte Stückzahl von 80 Stück gehalten worden.

Nachdem nach Ihren eigenen Angaben in der Zwischenzeit, abgesehen von der heutigen Meldung, von den damals in Reparatur befindlichen Ziehmaschinen zusätzlich reparierte zum Einsatz gekommen sind, müßte bei dem Mehr- oder Weniger-Taktverfahren, in dem Sie arbeiten, doch bereits eine Ausbringungssteigerung festzustellen sein.

Herr Ing. Mayer hat mir noch heute telefonisch nach Bekanntgabe der div. Maschinenausfälle gemeldet, daß trotzdem mit einer Tagesausbringung von 160 - 180 Stück zu rechnen ist, welche Stückzahl aber bis 17. - soweit reichen die Eintragungen in dem vorgelegten Programm, noch nie erreicht wurde.

Bezüglich des Zusammenbaues Schloß mit Griffstück ist gegenüber dem von Ihnen selbst erstellten Programm ab 13. bis inkl. 18. ein Rückstand von mindestens 550 Garnituren entstanden. Sie müssen trachten, aus dem festgestellten fertigen Umlauf von 2.000 Schloß und 1.100 Griffstücken in kürzester Zeit eine Aufholung, d.h. zusätzliche Lieferung zu Ihrer normalen Fertigungsausbringung durchzuführen.

Werksgruppe Waffen

(Fig. 57-1) Steyr directorate dated January 19, 1945.

English language synopsis: Mr. Brunner has shown me the delivery schedule and I have to conclude that since my last visit of December 13 no progress has been made in the production of the most important parts (frame and slide). I had specifically instructed you to report any production interruptions. Your notices/reports until now, including today's where you outline the breakdown of one of the frame manufacturing machines, do not account for the shortfalls. At the time of my visit daily slide production numbers were 120-140 and between the 10th and 12th 150. From the 13th a decline to 100, to the 16th with only 47 and the 17th none produced.

Further to conclude that nobody worked on slides on Sunday the 14th. Your production decline is not justified and I have to ask for a detailed report. Frame production took place on Sunday and according to the memo of January 13, 80 pieces were produced.
The added and repaired shaft operated machines should have shown an increase in production numbers. Mr. Mayer informed me today that even with the machine breakdown 160-180 pieces should be obtained, and these numbers have never been achieved. In regards to the program that you initiated on assembling slides and frames, there is a shortage of at least 550 units between the 13th and 18th.

You have to try to catch up with the established production plan and complete 200 slides and 1100 frames in the shortest order, which means adding these quantities to your set delivery figures. (Document courtesy of Waclaw Ustupski)

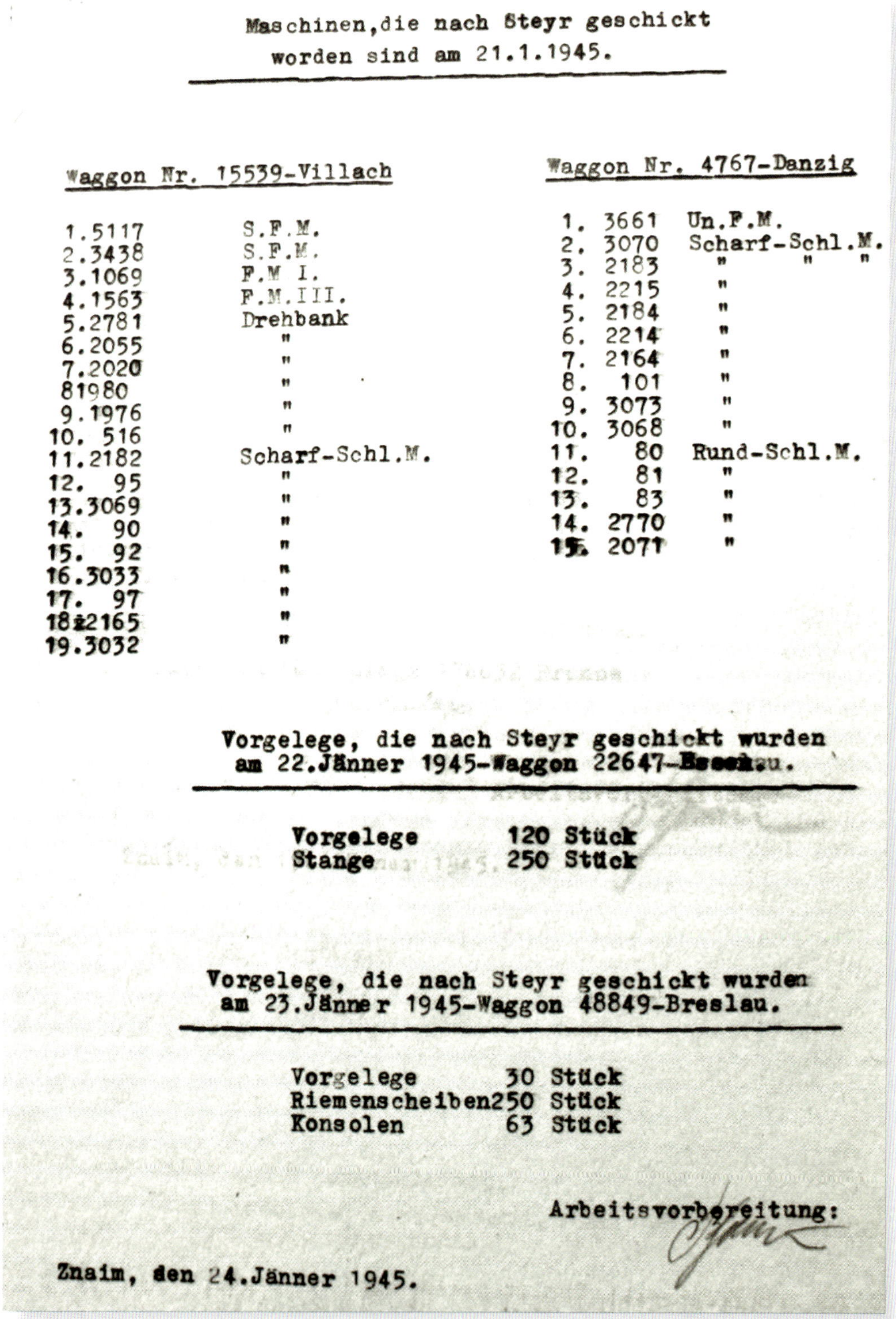

Maschinen, die nach Steyr geschickt worden sind am 21.1.1945.

Waggon Nr. 15539-Villach

1.	5117	S.F.M.
2.	3438	S.F.M.
3.	1069	F.M I.
4.	1563	F.M.III.
5.	2781	Drehbank
6.	2055	"
7.	2020	"
8	1980	"
9.	1976	"
10.	516	"
11.	2182	Scharf-Schl.M.
12.	95	"
13.	3069	"
14.	90	"
15.	92	"
16.	3033	"
17.	97	"
18.	2165	"
19.	3032	"

Waggon Nr. 4767-Danzig

1.	3661	Un.F.M.
2.	3070	Scharf-Schl.M.
3.	2183	" " "
4.	2215	"
5.	2184	"
6.	2214	"
7.	2164	"
8.	101	"
9.	3073	"
10.	3068	"
11.	80	Rund-Schl.M.
12.	81	"
13.	83	"
14.	2770	"
15.	2071	"

Vorgelege, die nach Steyr geschickt wurden am 22.Jänner 1945-Waggon 22647-Breslau.

Vorgelege	120 Stück
Stange	250 Stück

Vorgelege, die nach Steyr geschickt wurden am 23.Jänner 1945-Waggon 48849-Breslau.

Vorgelege	30 Stück
Riemenscheiben	250 Stück
Konsolen	63 Stück

Arbeitsvorbereitung:

Znaim, den 24.Jänner 1945.

(Fig. 58-1) Machinery Shipping Manifest – January 21, 1945

Notes: Although production lines were established, machinery was shipped back and fourth to Steyr and Gusen, possibly indicating changes in strategy where specific parts were to be made after Allied bombings. (Document courtesy of Waclaw Ustupski)

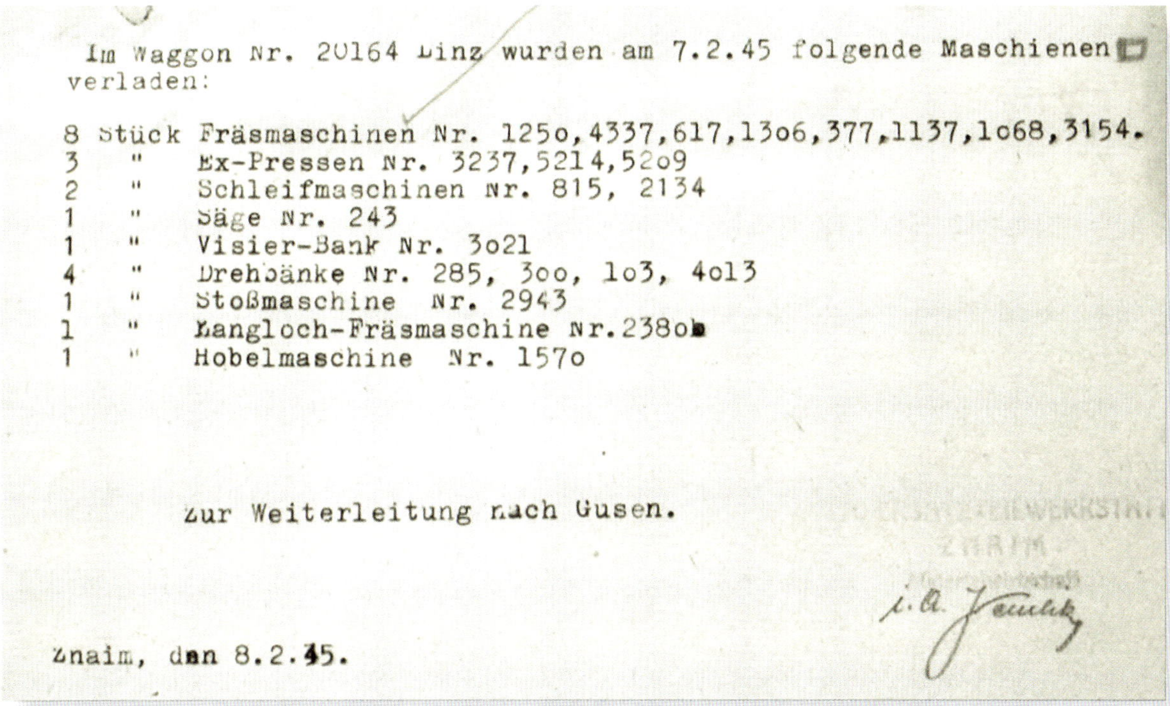

Im Waggon Nr. 20164 Linz wurden am 7.2.45 folgende Maschienen verladen:

8 Stück Fräsmaschinen Nr. 125o,4337,617,13o6,377,1137,1o68,3154.
3 " Ex-Pressen Nr. 3237,5214,52o9
2 " Schleifmaschinen Nr. 815, 2134
1 " Säge Nr. 243
1 " Visier-Bank Nr. 3o21
4 " Drehbänke Nr. 285, 3oo, 1o3, 4o13
1 " Stoßmaschine Nr. 2943
1 " Langloch-Fräsmaschine Nr.238o
1 " Hobelmaschine Nr. 157o

Zur Weiterleitung nach Gusen.

Znaim, den 8.2.45.

(Fig. 59-1) Machinery Shipping Manifest – February 7, 1945

Notes: Although production lines were established, machinery was shipped back and fourth, possibly indicating changes in strategy where specific parts were to be made after Allied bombings. Note that these machines are shipped to the concentration camp of Gusen. (Document courtesy of Waclaw Ustupski)

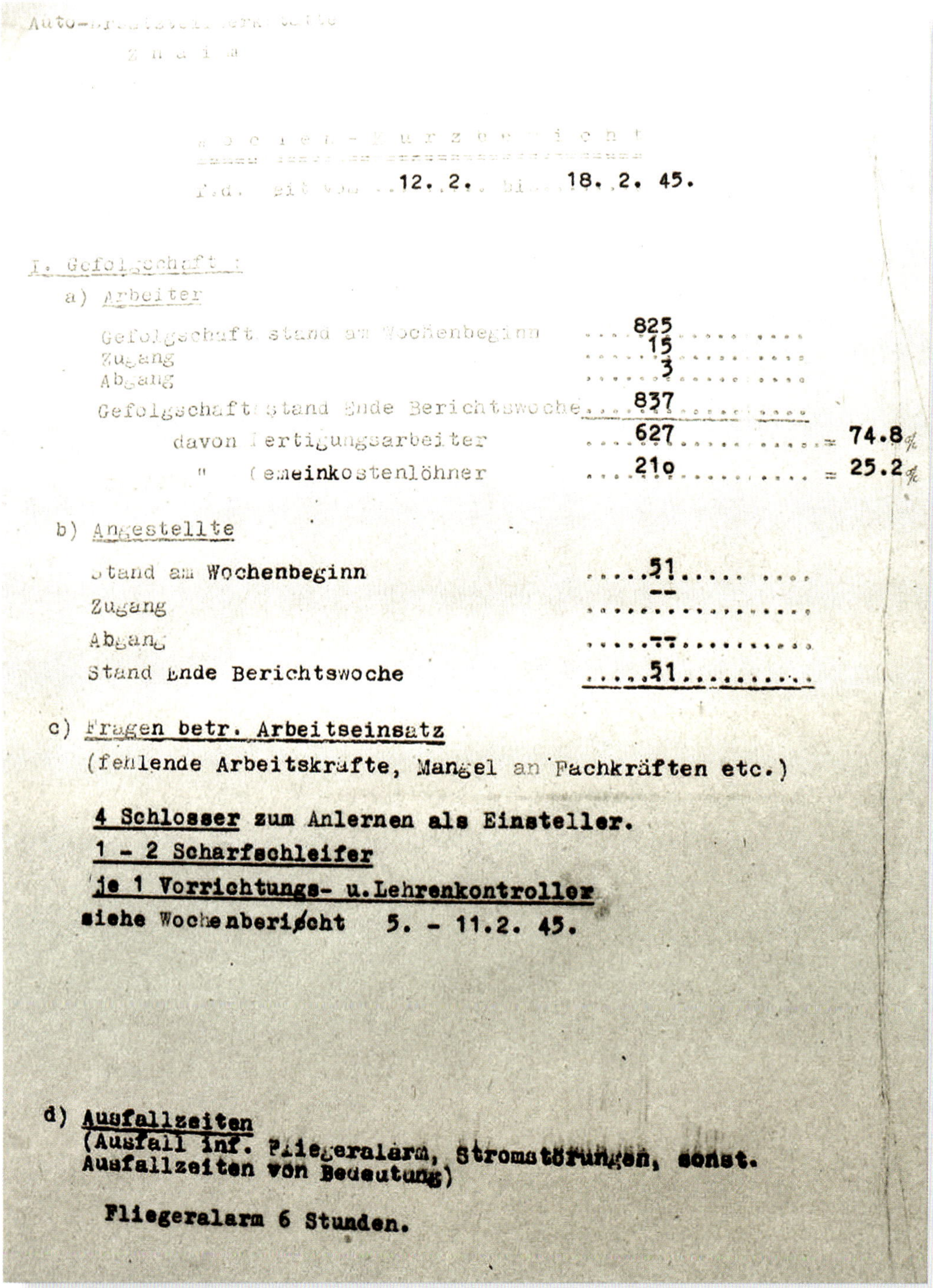

Auto-... Znaim

Wochen-Kurzbericht

f.d. Zeit vom 12.2. bis 18.2.45.

I. Gefolgschaft:

a) Arbeiter

Gefolgschaftsstand am Wochenbeginn	825	
Zugang	15	
Abgang	3	
Gefolgschaftsstand Ende Berichtswoche	837	
davon Fertigungsarbeiter	627	= 74.8 %
" Gemeinkostenlöhner	21o	= 25.2 %

b) Angestellte

Stand am Wochenbeginn	51
Zugang	--
Abgang	--
Stand Ende Berichtswoche	51

c) Fragen betr. Arbeitseinsatz
(fehlende Arbeitskräfte, Mangel an Fachkräften etc.)

4 Schlosser zum Anlernen als Einsteller.
1 - 2 Scharfschleifer
je 1 Vorrichtungs- u. Lehrenkontroller
siehe Wochenbericht 5. - 11.2. 45.

d) Ausfallzeiten
(Ausfall inf. Fliegeralarm, Stromstörungen, sonst. Ausfallzeiten von Bedeutung)

Fliegeralarm 6 Stunden.

(Fig. 60-1) Weekly workforce report from February 12 to February 18, 1945.

English language synopsis:
Worker count at beginning of week: 825
Added: 15
Subtracted: 3
Worker count at the end of the week: 837
From which are production (machinists) workers: 627
From which are basic workers: 210

Office workers: 51

Note the mention of disturbances: Air raids for six hours
(Document courtesy of Waclaw Ustupski)

.V. Versand :

SCA .

		Versand i. Ber. W.	Versand Jahresbeg bis Ber. W.
a) kpl. "Vis"- Sätze		400	4.500
b) montagefertige Teil			
705	Visier	565	565
708	Schlagbolzenve chluß	500	500
729	Stift f. Schaftv schlußstk.	--	--
730	Hahn	--	2.750
731	Hahnhebel	--	848
734	Hahnspanner	--	2.163
737	Spannungsunter scher	--	413
738	Abzug	--	--
743	Schloßhalter	602	1.554
744	Magazinsicheru	161	5.593
746	Druckfeder	--	--
754	Spannungslöser pl.	--	--
756	Feder f. Spann gslöser	--	6.099

WSCA. .

Versandschwierigkei n :

keine

Znaim, 6. März 1945. ....

Kaufmann. Leitung :

(Fig. 61-1) A late war Znaim shipment report dated March 6, 1945.

Notes: The reference to the 400 VIS Satze indicates "kits", assembled pistols without barrels and magazines, 4500 of such "kits' had been delivered. When pairing these numbers up to serial numbers collectors should keep in mind that there were most likely shipments that never reached Steyr due to Allied dominance of the skies and constant strafing of roads. This manifest was drafted two months before the Red Army entered Znaim. Note that this is the first indication of production of (rear) sights at Znaim. (Document courtesy of W. Ustupski)

A Visual Representation of the Steyr Znaim Complex after Allied Bombing

(Fig. 62-1) The Znaim factory was initially bombed in January, 1945.
(Photographs courtesy of W. Ustupski)

(Fig. 63-1) The upper left photograph shows destroyed commodes left over from the Znaim factory's porcelain business. (Photographs courtesy of W. Ustupski)

(Fig. 64-1) More views of Znaim factory destruction. (Photographs courtesy of W. Ustupski)

Chapter V
VIS Markings

Markings on Polish Production

Polish production VIS pistols are often called "Eagles", and for a good reason. The left side of their slide carries a legend with a large crowned Polish eagle in its center. To the eagle's left are the maker's name and the date of production. To the right are the model designation and Polish patent number. There is another, small Polish eagle on the slide to the left of the hammer drop lever - this is the Polish firing proof, and signifies that the slide has passed a proof test with elevated pressure ammunition. A similar small Polish eagle, also a firing proof mark, is found on the left side of the barrel lug.

The prewar VIS is also marked with an oval containing a capital letter over a number, on the upper left web of the trigger guard. This is a final inspection mark, indicating acceptance into service. The front lip of most prewar VIS magazine bottoms is also so marked. Prewar pistols also show a considerable number of various assemblers' and sub-inspectors' marks, which can be geometric, mathematical, or musical in inspiration. They can be an interesting field of study in themselves. These markings are found on most components of the weapon, Finally, the barrel is marked with its exact metric bore diameter.

(Fig. 65-1) Final inspection marking from Major Dzierzynski

(Fig. 65-2) Final inspection marking from unknown inspector

(Fig. 65-3) Final inspection marking from Captain Czuma

(Fig. 65-4) Final inspection marking from Captain Gadomski

(Fig. 65-5) The left side of a prewar VIS barrel lug, showing the Polish firing proof and bore diameter marking. (Photograph by Anthony Vanderlinden)

(Fig. 65-6) The firing proof on a prewar VIS slide. (Photograph by Anthony Vanderlinden)

(Fig. 65-7) The left side slide marking of a 1936 VIS. (Photograph by Anthony Vanderlinden)

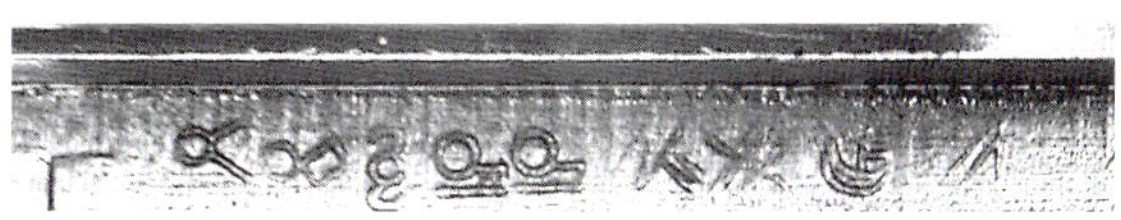

(Fig. 65-8) View of workmen's marks inside the slide bottom of a prewar VIS. (Photograph by Anthony Vanderlinden)

Markings on German Production

The most immediately apparent difference in the marking of wartime, German produced VIS pistols is the legend on the left side of the slide. The Polish eagle has been deleted and the legend is now in a single line, giving the maker's name, the model, and the patent number. The small Polish eagle proof marks on slide and barrel have now been replaced with small eagle-and-swastika stamps, which are the Nazi German firing proof and acceptance markings.

Early production German made pistols will also have a stamped marking on the left side of the slide giving the German designation of the VIS- "P.35(p)".The acceptance marks of the local branches of the German Army Weapons Office (the Heereswaffenamt) for both Radom and Steyr will also be found in various forms and places on these pistols, throughout wartime production. Very late pistols manufactured after the evacuation of Radom will only show the Steyr mark. German production pistols also show various assemblers' and sub-inspectors' marks, but they are simpler and fewer than those on prewar guns, especially so on late war guns. These various markings are illustrated in the photo gallery, as are the markings on wartime variations such as German Navy and BNZ marked pistols.

When shipping kits from Radom and Znaim, the Germans employed an inventory numbering system which was applied to the frame and slide. This inventory numbering system had no relationship to the gun's serial number and inventory numbers are not matching on the frame and slide of completed pistols.

(Fig. 66-1) Inventory numbers on late war VIS slide and frame. (Photographs by Anthony Vanderlinden)

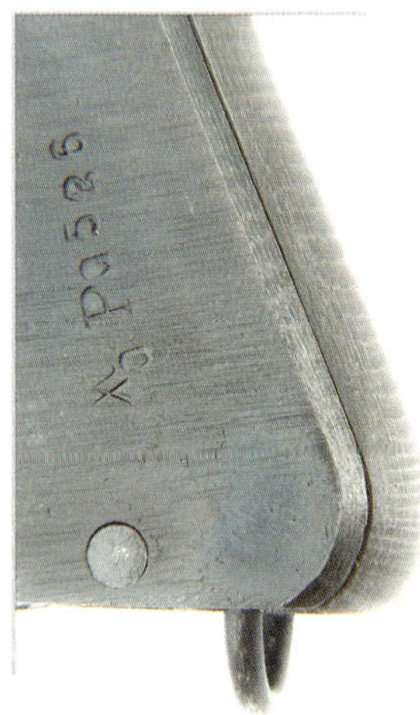

Details on Waffenamts

A Heeres waffenamt number was not assigned to any specific factory. Instead a specific number was assigned to an acceptance/ Inspection Officer. This Inspection Officer maintained the number wherever he was commissioned to work. Although we do not know the name of the accepting officer working at Radom (WaA77) we do know some of his background history.

An Inspection Officer oversaw an inspection team, these teams varied in size according to the needs and size of operations they were to oversee. A team characteristically constituted of enlisted men (engineers and experienced manufacturers), NCOs and clerical staff. The Inspection Officer worked with the team of the facility where he was assigned, teams remained at a specific facility and did not transfer with the Inspection Officer. The Inspection Officer did not personally apply the markings to the parts, his team was issued a number of dies which were replaced when needed.

By 1940 the Radom factory inspector most likely had the title of Technical Inspector First Class with the equivalent military rank of Hauptman (captain) as he had been in service for more than five years. Some of his background work included inspection of bayonets in Solingen starting in 1935 and inspecting K98k parts at Erma starting in 1937. Aside from being responsible for accepting VIS parts at the Radom plant, his team also accepted leather goods (including holsters - but not VIS holsters) from the Kromolowski & Sohne Lederwaren Fabrik in Radom as well as holsters made by Arba GMBH in Petrikau (now Piotrkow Trybunalski) Poland, some 113 kilometers (70 miles) from Radom.

Radom was the center of one of the 14 regional armament inspectorates in the entire German system. It is therefore not unusual that the same team managed more than one factory, although both leather goods manufacturers fell under their control only late in the war in 1944. Interestingly enough, most armaments inspectorates were based in the capital cities of occupied countries, Radom was an exception. Understanding the sequence and acceptance markings on the VIS pistol proves to be extremely challenging as we do not know the methodology used by this Germans for the VIS pistol. All procedural documents pertaining to acceptance and methodology were destroyed by Allied bombing of the headquarters in Wunsdorf in the Spring of 1944.

Production acceptance of parts:

One important aspect that is often overlooked is that the Eagle (E/77, E/623, E/189) marking is not a simplified derivative of the Heereswaffenamt (WaA77, WaA623). In all wartime German controlled plants, the Eagle marking is applied only to parts that are furnished to another assembler/manufacturing plant. This indicates that the part fit set standards but that the fitting, assembly, fire proofing and final acceptance was to be done in another plant/location. As a general rule, subcontracters used Eagle markings and not WaA acceptance markings.

The waffenamt marking was traditionally (exceptions do exist) applied to parts and complete guns that were built internally where they are accepted.

As such there were numerous plants that used both WaA markings as well as Eagle markings simultaneously: WaA markings on internal produced guns and Eagle markings on unfinished parts that were shipped on to other factories and proofing facilities.

Final acceptance and firing proof:

The final acceptance marking was applied only after the pistol was marked with the firing proof. One German practice was to stamp the firing proof and final acceptance next to the pistol's slide legend.

Traditionally this means a WaA marking stamped close to or next the firing proof on guns that are built internally (all major components). Guns assembled from major parts originating from other plants were accepted with an Eagle marking of the plant were they underwent the firing proof. Here too the Eagle marking is stamped near or next to the firing proof.

Some cases can have two or multiple identical Eagle markings on the slide near the firing proof, this indicates that the slide was built in the same facility as where the pistol underwent final acceptance but that the frame and/or barrel originated from another factory. This is often encountered on P-38 production.

Accepting the VIS:

Very early occupation pistols were entirely manufactured at Radom. This included barrels, magazines, assembly, proofing and acceptance. Those early guns were all WaA77 marked near the firing proof. Shortly afterwards, the Radom production was limited to producing kits while barrels were made at Steyr, but the WaA marking continued to be applied instead of the Eagle 77 marking. This was not unusual. From other plants we know that some markings continued to stay in service unless the inspection officer was replaced with another inspection officer (with his own specific number).

Pistol N° 1278 (fig.147-1) is a perfect example of pistol that was originally intended to be accepted at Radom. The WaA marking is present on the slide, in the location where the firing proof would be stamped to the right, followed by the final acceptance (WaA) to the right of the firing proof. Instead this was shipped to Steyr where the firing proof was applied and E/623 marking (E/623 instead of WaA623 as the parts had been produced outside of Steyr).

While examining pistol N° 8441 (fig.148-1) it becomes apparent that the Germans relocated the marking to the appropriate location on the slide for the E/77 marking but did not change the die. This may be attributed to many reasons including that no E/77 die was available. Technically the marking should have been changed to E/77 but the location of the stamping was correct, it was certainly a low priority.

Understanding late war acceptance is more complicated as we do not know where the pistols were proofed and accepted. These could have been proofed at Steyr or one of the many Steyr sub-plants. We know that the Znaim sub-plant used the E/623 marking and it is safe to assume that the Gusen camp also used the E/623 marking. From documents we know that kits without barrels were shipped to Steyr from Znaim, so Znaim was not a location of final proofing and acceptance.

For unknown reasons the Germans started using WaA623 on much of the second alphabet series, but exceptions of E/623 do exist. The WaA623 marking shows up on Steyr manufactured frames and slides but also on older Radom produced parts assembled at Steyr. Additionally they reverted back to using just E/623 on late war pieces in the "K" serial block range. These irregular approaches can be attributed to the late war desperate situation where Allied bombing crippled much of the facilities and manufacturing and other services were relocated and merged with remaining operating facilities.

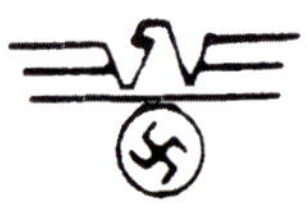

(Fig. 68-1) German firing proof, indicates that the pistol passed a proof test with elevated pressure ammunition.

(Fig. 68-2) Used at Radom on early completed and accepted pistols and on parts kits supplied to Steyr.

(Fig. 68-3) Used at Radom on parts and pistol kits.

(Fig. 68-4) Final acceptance marking applied at Steyr or one of Steyr's sub-plants. This marking is found next to the firing proof.

(Fig. 68-5) Used at Steyr and Steyr sub-plants (including Znaim) on parts and pistol kits, also used as final acceptance marking next to firing proof.

(Fig. 68-6) This marking only applies to magazines manufactured at one of the Steyr sub-plants in Warsaw. Most likely this facility supplemented magazine production as the factory was more specialized in stamped steel production. It manufactured MG42 components, Bergmann MP35 components and VIS magazines.

Chapter VI
VIS Serial Numbers

It is evident that the application of final serial numbers to the profiled frame, slide, grip safety, mainspring housing, barrel, hammer drop lever, extractor, firing pin stop, hammer and trigger were done prior to the application of the pistol's final blue finish, while the parts were still in-the-white. Perhaps this applies to the application of proof marks as well. Prior to this time, fitters and sub-inspectors had added their marks to the parts and assemblies they had worked on. After finishing, the numbered parts were reunited and the completed, matched pistol was subjected to final inspection. Once passed, pistols were greased, boxed individually, and shipped.

Polish Production Serial Numbers

Polish production: Total prewar production is generally stated to have been approximately 49,400 pistols (the latest Radom factory historical material gives a number of 51,000), with mismatched parts in the last 2,000.

According to Still, annual production was as follows;

1936: numbers 01-5500
1937: numbers 5500-8800,
1938: numbers 8800-29000,
1939: numbers 29000-49400.

According to contributors' reports and observations:

1936: 01 through 5360
1937: 5359 - 8807
1938: 8808 through 27450
1939: 27451 through 49014

Note that 1937 was the lowest production year, due to Poland's arms industry's being busy filling rifle orders from Spain, then engulfed in civil war. This, according to current factory materials led to a cessation of pistol production at the end of the 1936 fiscal year, in March, 1937.

Some pistols with 1937 dated slides have been reported outside of the above given serial number range, indicating that these ranges should be considered general guidelines, with exceptions to be found. However, due to the rarity and collector value of 1937 pistols, counterfeits may exist. and expert advice should be sought if a purchase of a weapon outside the usual range is contemplated. The "mismatched" pistols at the end of Polish production do, as an aside, show that the Poles had by this time achieved a fair degree of parts interchangeability. This is significant when one contemplates the relatively tiny production numbers of the prewar VIS with, for example, Colt M1911 and 1911A1 production numbers.

Pistols in the Polish Army: 1936-1939

III On October 1, 1936, the Polish Army reported being equipped with the following pistols:

7.65mm Pistols (no break down as to type):	9660
8mm Roth-Steyr:	168
9mm VIS trial series:	40

IV. As of January 7, 1938, the official numbers were:

7.65mm pistols:	9658
9mm pistols:	1481
8mm Roth-Steyr:	218
9mm VIS M35:	29,416

As it appears, rearmament with the VIS was progressing well by 1938, but had not yet displaced older pistol types. Police and border guards would still have been armed with Nagant amd French revolvers, and officers would have retained their self-purchased handguns as well. The rifle situation had been even more chaotic in the immediate aftermath of Poland's 1918 rebirth, but was also made rational by 1939.

(Fig. 70-1) The above figures are from BRON STRZELECKA I SPRZET ARTYLERYSKI FORMACJI POLSKICH I WOJSKA POLSKIEGO W LATACH 1914-1939 , by Andrzej Konstankiewicz, Lublin, Uniwersitet Marii Curie-Sklodowski, 2003; and BRON STRZELECKA WOJSK POLSKICH W LATACH 1717-1945, by Marian Maciejewski, Szczecin, Wydawnictwo Glob, 1991.

German Production Serial Numbers

German production: Initial German production, utilizing leftover Polish parts and mostly Steyr-made barrels, was serial numbered from 001 to about 12,000, including a "zero" series with a "0" prior to a number of usually three digits, and a "regular" series beginning at "1". Thereafter, serial numbering was in groups of 10,000 (actually 9999) pistols, each with an alphabetical prefix running from "A" through "Z", and again through the letter "K" (no pistols from a second series "L" have been reported). Not all letters of the alphabet were used, notably the letters "I', "O", "Q", "V", "X" and "Y". Total German production has been estimated at 312,000 pistols (up to 350,000 in current Radom factory materials).

Initial German production numbered parts in the Polish fashion; later shortcuts in manufacturing eliminated numbers on extractors and firing pin retainers, reducing the numbered parts from nine to seven. The serial number of the magazine safety also becomes visible on late wartime wartime pistols, the number is most often applied to the left side, although right side numbering was also done.
Late war second alphabet series "K" pistols, have their slide serial number under the slide and on top of the right slide rail. Previous VIS production had slide numbers under the slide and inside the left side milling cut. This change in numbering practice coincides with new slide production at Steyr/Znaim.

(Fig. 70-2) Comparison of serial numbering under the slide of Radom (left) and Steyr (right) production. (Photographs by Anthony Vanderlinden)

Note on BNZ Serial Numbers

"BNZ" pistols have been reported with serial numbers considerably off from the "K1440-K2400" range reported by Still. Reported numbers range from just under K1000 to over K2400. This makes some sense, considering the chaotic conditions of their manufacture. It certainly seems to confirm the "top of the pile" assembly of components at war's end, at least. Needless to say, this makes the fraudulent matching of a "BNZ" slide with any second K series frame and barrel all the easier. Accordingly, pistols with "off" numbers should be carefully examined due to the presence of counterfeits.

(Fig. 71-1) A German armourer works on various pistols including the VIS. (Photograph courtesy of Waclaw Ustupski)

Chapter VII
VIS Finishes

Rust Blue

A high-quality, deep rust blue finish was applied to all prewar production.

Wartime Salt Blue

The time consuming rust blue process was replaced by the faster salt blue process early in the war. Additionally the appearance from a gloss blue finish to a duller finish took place in the first "K" serial block when the Germans cut back on metal polishing, affecting the overall appearance of the pistol.

Phosphate

The VIS pistol, as it was produced at Steyr/Znaim in 1945 was a crude but utilitarian weapon with phosphate finish on its frame and slide, a bright steel barrel, and a blued finish on a number of parts (the hammer and sometimes the hammer drop lever, the slide stop, the magazine catch, the extractor, the rear sight, and sometimes the grip screws). The lattter parts are finished in a rather thin blue which contrasts pleasingly with slide and frame.

Several questions have arisen with regard to pistols of earlier alphabet serial number series which are also phosphate finished on slide and frame, and are also fitted with grooved wood grips. Such pistols are sometimes found in any of the second alphabet serial number ranges, but a fair number of them exist in the second "A", "B" and the second "H" series. The questions asked are when and where were these late war pistols produced.

On inspection, the pre-second "K" series pistols with second series "K" characteristics have been found to be serialized inside the left milling cut of the slide, most or all have E/77 markings, and most are fitted with flexible recoil spring guides.

Accordingly, it is the author's opinion that these phosphated pistols are probably from a rather large group of Radom produced kits which were not or could not be assembled into complete pistols prior to 1945. This is plausible because the Steyr plant was increasingly subjected to Allied bombing raids during the later war years, sustaining considerable damage in February of 1944. This attack certainly disrupted operations at Steyr, and led to the relocation of Steyr rifle and machine gun production to underground facilities associated with the Gusen concentration camp.

(Fig. 72-1) A German soldier of Asian descent poses with his VIS for a studio portrait. (Photographs courtesy of W. Ustupski)

The author concludes that all or practically all VIS pistols with grooved wood grips and phosphated slides and frames were assembled in 1945, contemporaneously with "K" series Steyr produced VIS and BNZ marked pistols.

Chapter VIII
Variations of the VIS
.22, .45 and Presentation Pistols

The VIS was produced in caliber 9mm Parabellum. A .22 caliber version exists in a museum in Budapest, Hungary. Upon inspection it does not appear to be a factory product.

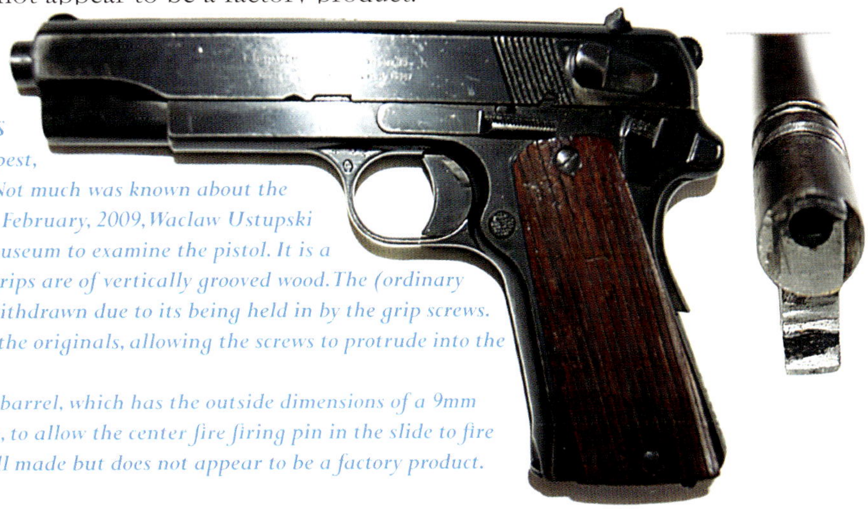

(Fig. 73-1) Illustrated is the .22 caliber VIS in the Hungarian Army Museum in Budapest, which has been shown in previous books. Not much was known about the particulars of this pistol until recently. In February, 2009, Waclaw Ustupski obtained permission from the Budapest museum to examine the pistol. It is a 1937 Polish VIS, serial number 4887. The grips are of vertically grooved wood. The (ordinary eight shot 9mm) magazine could not be withdrawn due to its being held in by the grip screws. (The replacement grips are thinner than the originals, allowing the screws to protrude into the magazine well).
The slide was unmodified. The .22 caliber barrel, which has the outside dimensions of a 9mm barrel, was bored off center toward its top, to allow the center fire firing pin in the slide to fire the rimfire .22 cartridge. The barrel is well made but does not appear to be a factory product. (Photographs by Waclaw Ustupski)

Factory records indicate that two VIS pistols in caliber .45ACP, together with shoulder-stocks, were sent to Argentina for military trials in 1937, but the whereabouts of these weapons is not known. Polish published sources name Jerzy Podsendkowski of the Warsaw State Rifle Factory's Bureau of Studies as the developer of the .45 VIS. Specimens could also have been experimentally produced in other calibers, the 7.65MM Parabellum comes to mind, but their existence is only speculative.

(Fig. 73-2) The right side of 1936 Presentation VIS N° 0034. The markings usually on the left side have been moved to the right side to make room for an inscription on the left side. The author has not been able to examine this pistol and can not attest to its authenticity. (Photograph courtesy of Rock Island Auction Company)

A series of presentation pistols were produced for Polish dignitaries, with the usual slide inscription moved to the right side of the slide, and an inscription in gold naming the recipient on the left. Pistols given to the President of Poland, Ignacy Moscicki, and to the Defense Minister, Tadeusz Kasprzycki, are known to exist. There may have been as many as 50 such presentation arms.

(Fig. 74-1) "The left side of VIS Nº 0034" with an inscription to " Minister of Defense, General of Division Tadeusz Kasprzycki". The author has not been able to examine this pistol and can not attest to its authenticity. (Photograph courtesy of Rock Island Auction Company)

The Full Automatic Version

There were some prewar experiments with a full-automatic version of the VIS with an extended magazine and shoulder-stock. The experiments revived the possible use of the shoulder-stock. The adoption of the MORS sub-machinegun in 1939 put an end to the full automatic consideration. It never went into production and none are known to exist.

The 1992 Replica VIS

After a hiatus of 53 years, the "Lucznik"("Archer" in Polish) factory in Radom, which is the successor corporation to the former Radom arms factory, decided to produce a duplicate copy of the VIS for the benefit of collectors. The slide legend on these pistols is the same as that on prewar pistols, with a somewhat different Polish eagle. Serial numbering began at A00001 for a projected run of some 1000 pistols, to be sold for U.S.$1000, to include a wooden case and replica accessories. These pistols have wooden grips with the "FB" and "VIS" markings, They also differ from the originals in their sight picture, featuring a square rear sight notch and a matching front sight profile. The originals have a "V" notch at the rear and tapered front sights. Pistol number A00024 from this series is illustrated. Only 26 were actually produced.

At the 2011 SHOT show in Las Vegas, Nevada, the Radom factory unveiled a new production of the VIS, with the first two examples on display. These weapons are to be produced by computerized CNC machinery, and could be sold for a competitive price, if a market for them is found to materialize.

(Fig. 75-1) VIS pistol, N° 24, produced in 1997. One of allegedly 26 produced. Note that the pistol lacks internal markings.
(Photographs by A. Vanderlinden, courtesy of Ron Pinkenburg)

(Fig. 76-1) Right side and boxed views of 1997 VIS N° A00024.
(Photographs by A.Vanderlinden, courtesy of Ron Pinkenburg)

Polish Underground VIS Pistols

As we mentioned above, the manufacture of barrels and final assembly of wartime VIS pistols were done at Steyr. This did not prevent the Polish workers at the Radom factory from stealing the parts they unwillingly produced, to provide pistols to the Polish underground, and although they are rare, underground VIS pistols may be found in a few collections (see pages 227-231). None have serial numbers, although the late war slides and frames used on examples we have inspected do have the inventory numbers used by the Germans to prevent pilferage. Obviously the Poles found a way around the Germans' inventory system, such as by double stamping the same number on two slides or frames. Three underground pistols are illustrated in our photo gallery. One has been fitted with a prewar VIS barrel, doubtless the result of scrounging for an unavailable part. The other is fitted with an otherwise unmarked barrel which has the Nazi German firing proof on the left side of its underlug. This may be a stolen barrel, but in the author's view it is more likely an underground-produced barrel which was so marked in order to confuse the Germans if the pistol were to fall into their hands. According to Gwozdz, the Polish underground organized the production of Vis barrels in Warsaw. The first of these barrelmaking shops was located on Terespolski street. After its liquidation, barrelmaking for the underground VIS was moved to N° 18 Leszno street, where production was supervised by the engineers R. Bialostocki and T. Czaikowski. Production at this second location continued until the 1944 Warsaw Uprising.

(Fig. 77-1) Polish resistance fighters during the Warsaw Uprising: The fighter on the left is holding a VIS, the resistance fighters can be identified by their armbands. (Photograph courtesy of Waclaw Ustupski)

(Fig. 77-2) A Polish resistance fighter holding two VIS pistols. (Photograph courtesy of Waclaw Ustupski)

(Fig. 78-1) Polish members of the Underground and resistance fighters with VIS pistols. (Photographs courtesy of Waclaw Ustupski)

(Fig. 78-2) Rare photographs of the Warsaw Uprising: Polish resistance fighters with VIS pistols. The organizational skills of the Polish people during the uprising were staggering, women and children participated in all aspects including combat. (Photographs courtesy of Waclaw Ustupski)

Chapter IX
Accessories
Polish Holsters

The prewar VIS holster is of a "hardshell" type, with compartments for the pistol, two spare magazines, and a cleaning rod. This holster has flaps which cover the top of its main and magazine compartments, which flaps are secured by metal studs when closed. There is a belt loop in the holster's rear, as well as two metal rings for a shoulder strap. At least three Polish leatherworking firms manufactured these holsters. Some of these holsters will be found with makers' marks, as well as with the Polish oval letter-and-number acceptance mark. The prewar holster was also used unmodified by the Germans and even Soviets during and after the war. The holsters were originally brown in color, but many were dyed black by the occupying Germans. Holsters used by the Polish armored troops were also dyed black.

Collectors should be aware that reproductions of the prewar VIS holster have recently been produced in Poland, some apparently on the original blocks. If properly "aged", they could easily pass for the real thing. Caveat Emptor!

(Fig. 79-1) Prewar holster These are of the European "hard shell" type, and are rather rare. Originally brown, the Germans dyed a number of them black. The holster has a compartment for two spare magazines, and a place for a cleaning rod. The maker name and address was ink stamped on the inside of the holster's flap, and is sometimes still legible. (Photographs by A. Vanderlinden, courtesy of Waclaw Ustupski)

(Fig. 80-1) Prewar holster with prewar Polish officer's belt. The eagle maker's stamp is either green or purple in color. (Photographs by Anthony Vanderlinden.)

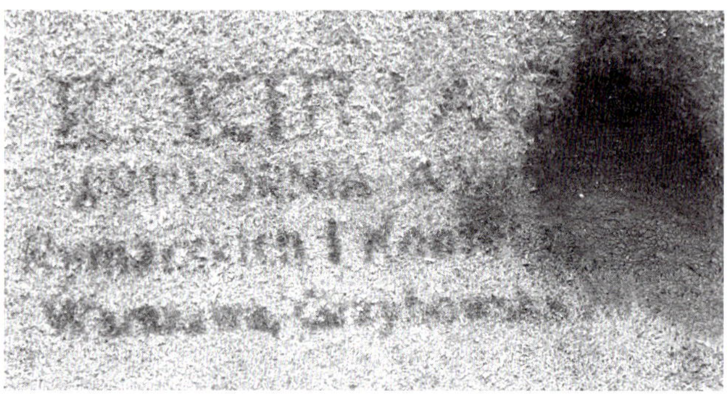

(Fig. 80-2) Variances of makers' markings. Top: K. Kirjacki, Leather Products Maker to Civilian & Military - 43 Grzybowska St. Warsaw. Right: Unidentified maker in Warsaw (Photographs by Anthony Vanderlinden, courtesy of Robert Burden and Francis Kennedy)

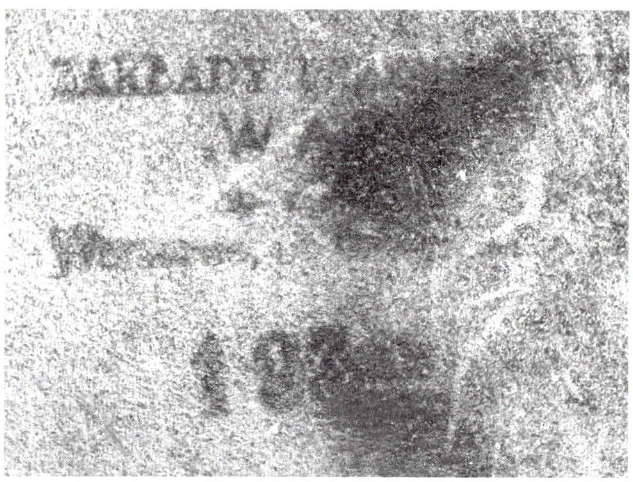

(Fig. 81-1) Prewar holster with rarely encountered shoulder-straps. The holster is date stamped "1937", note the purple eagle maker's stamp. The inset photograph shows the correct hardware for the shoulder straps. (Photographs by Anthony Vanderlinden, courtesy of Waclaw Ustupski)

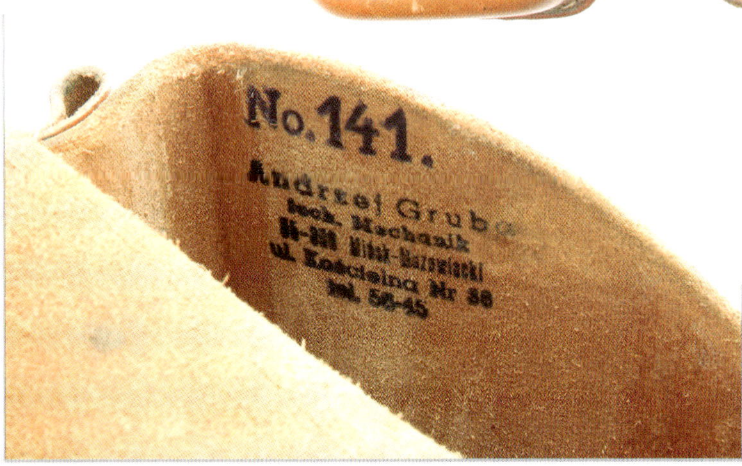

(Fig. 82-1) Replica of prewar holster. Some have been artificially aged to appear original. Caveat Emptor! Note the modern telephone number in the maker's information stamped inside this example.
(Photographs by A.Vanderlinden, courtesy of Waclaw Ust-upski)

German Holsters

The German holster for the VIS is of much simpler construction. It is a black or brown pistol shaped holster featuring a protective flap, and a pouch for one spare magazine. This holster has two belt loops in the rear, and a closing strap and metal stud in front. Two variants exits; the standard holster with pull-strap and late war simplified version without pull-strap. Many are marked with the Steyr "BNZ" maker's code, but other makers' marks are also to be found. Modern copies of this holster are also around, and the collector should beware of counterfeits here as well.

Many wartime holsters were modified to be attached to U.S. pistol belts. Many were modified by servicemen but also at the United States Pirmasens army depot in Western Germany, some were also so modified in Norway post-World War II. The modified holsters and VIS pistols were issued to depot guards, many of whom had been in the Polish armed forces and chose not to return to Communist Poland. Pirmasens is currently the location of the Westwall museum.

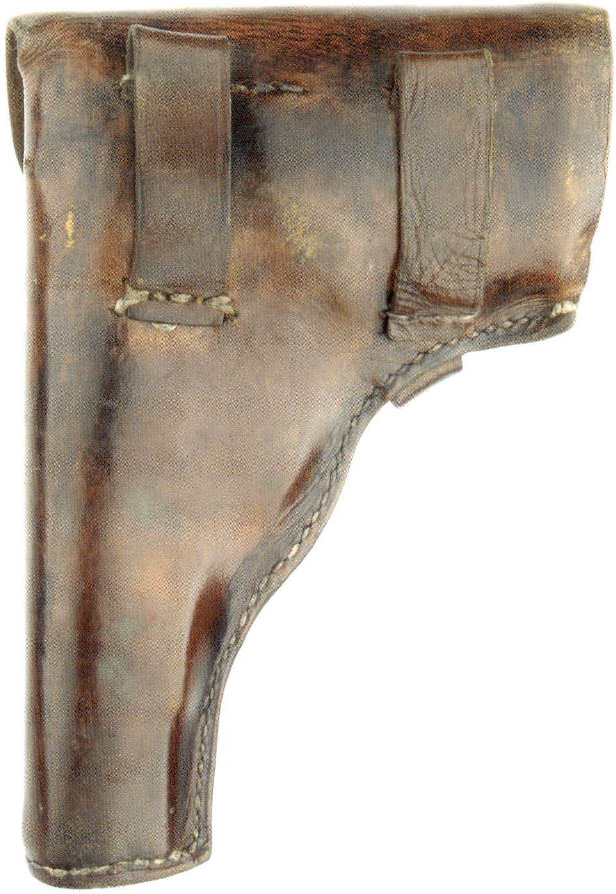

(Fig. 83-1) Wartime holster variant with pull-strap:
These may be brown or black in color. They will rarely show German Navy markings (beware of counterfeits). The majority were manufactured by the Steyr factory (wartime code bnz). One wartime holster has been observed with the " epf" manufacturer's code, indicating production at Schwientochlowitz, in Silesia now Swietochlowice in Poland. This was the site of a wartime concentration camp as well as a postwar prison/labor camp. Most observed holsters with this coding have been FN Browning High Power or P38 holsters. (Photographs by Anthony Vanderlinden, courtesy of Waclaw Ustupski)

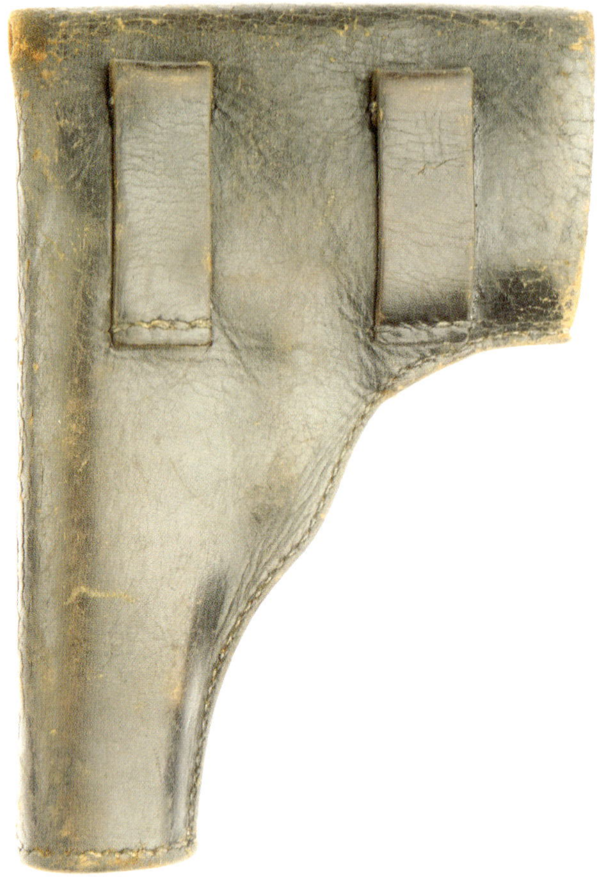

(Fig. 84-1) Late war holster variant without pull-strap:
These may be brown or black in color. The pull strap was eliminated from production as part of manufacturing simplifications. The holster is marked "GRZ" for Kruger in Breslau, some were WaA182 marked. Note the "194" marking with missing year digit, this is often encountered on last ditch manufacture. Note also the missing leather protection patch over the closure hardware. (Photographs by A.Vanderlinden, courtesy of Waclaw Ustupski)

(Fig. 84-2) The original German design for the holster provided ample protection but the high holster wall made the pistol difficult to remove. This was resolved by adding a pull-strap that rises the pistol out of the holster. Late wartime holsters were simplified and lack the pull-strap making it difficult to remove the pistol, another sign of simplified production in desperate times. (Photograph by Anthony Vanderlinden)

(Fig. 85-1) Top: A German soldier poses with his VIS holster for this studio portrait. Right: A German soldier climbs on the Polar Circle marker in Saltfjellet, Norway. Note that he is equipped with a prewar Polish VIS holster. (Photographs courtesy of Waclaw Ustupski)

(Fig. 85-2) Maker variant: Note the "5" digit, a simplified marking for 1945. The leather protection patch over the closure hardware is retained with this manufacturer. (Photographs by A.Vanderlinden, courtesy of Francis Kennedy)

(Fig. 85-3) Maker variant: "GRZ" for Kruger in Breslau, this one is WaA182 marked (compare with fig. 54-1). Note the "194" marking with the missing year digit and the missing leather protection patch over the closure hardware. (Photographs by A.Vanderlinden, courtesy of Francis Kennedy)

(Fig. 86-1) German shoulder holsters are typically unmarked, these can be identified by carefully examining wear markings in the leather for both pistol and magazine, the spare magazine pouch is often a prominent indicator if the holster was used with the VIS or not. Left: A German shoulder-holster. Right: A variant with chest strap. Many variances exist as no patterns were standardized. (Photographs by A.Vanderlinden, courtesy of Waclaw Ustupski)

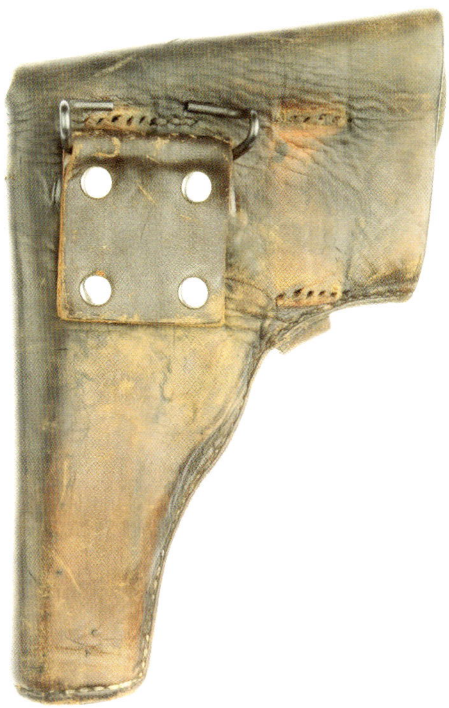

(Fig. 87-1) A U.S. GI poses with his VIS and holster for a studio portrait. The holster was modified to fit his issue belt. Note that the pistol appears to be nickel plated yet careful examination leaves it undetermined if the pistol is just reflecting studio lights or if it was indeed nickel plated. (Photograph courtesy of Anthony Vanderlinden)

(Fig. 87-2) Right: The U.S. modified German holsters at the Pirmasens arsenal. Some were also so modified in Norway post-World War II. (Photographs by Anthony Vanderlinden, author's collection)

Cleaning Rod

The Polish cleaning rod is made of brass, with a (often invisibly) brazed-on circular handle, and a flattened front section slotted toward its front end for a cleaning patch. It is found marked with the Polish oval acceptance stamp, and may also show what is a probable maker's mark.

There is no specific German cleaning rod for the VIS, due to the fact that the World War II German Army employed its common R. G. 34 "tobacco can" cleaning kit for both rifles and 9mm issue pistols. See Gotz and Sturgess, PISTOLE PARABELLUM, pp. 1524-6 (Collector Grade, 2010).

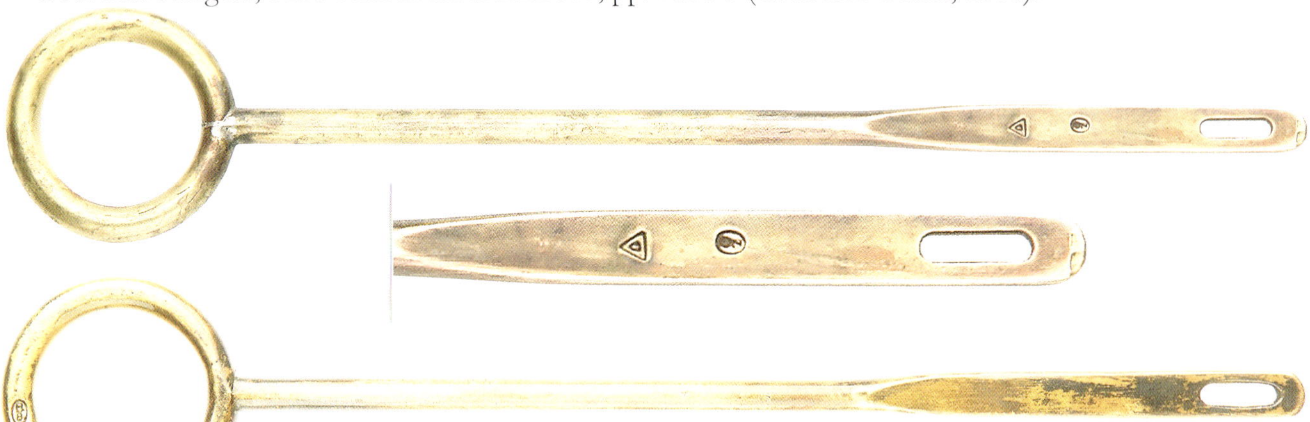

(Fig. 88-1) Prewar cleaning rods: The attachment of its loop and the flat surface at the end are distinguishing traits. They are stamped with an inspector's mark in an oval, two variations are shown. While most are of brass, some white colored ones are known. They may simply be nickel plated. The Germans did not produce a cleaning rod specifically for the VIS.
(Photographs by A. Vanderlinden, courtesy of Waclaw Ustupski)

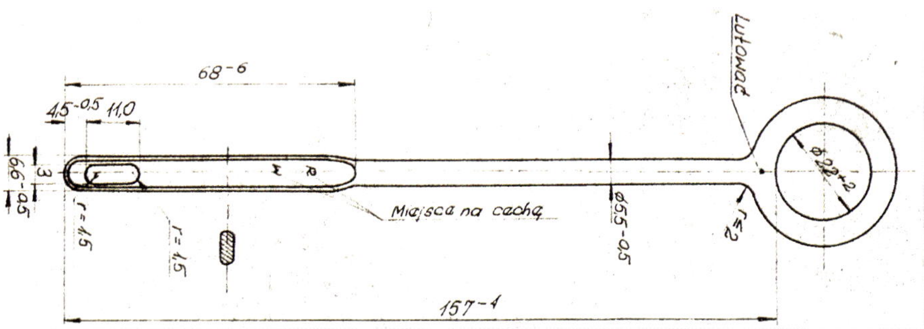

(Fig. 88-2) Original drawing for the VIS cleaning rod.
(Courtesy of Waclaw Ustupski)

(Fig. 88-3) German World War II R.G.34 cleaning kit for rifle and pistol use.
(Photographs by A. Vanderlinden, from the M. Peter Kuck collection)

Lanyard

The Polish armed forces issued a braided leather lanyard for the VIS. Due to their fragility, originals are extremely rare. Modern copies have been made, but they are also uncommon.

Shoulder-Stock

Prewar and early wartime VIS pistols posses a mounting groove milled into the mainspring housing, for the attachment of a shoulder-stock. One confusing feature of this is that two different widths of groove exist, with no correlation between serial number of pistol and width of groove. While it is known that a shoulder-stock was proposed for the VIS, and that a few were manufactured before the war, originals appear to be nonexistent. In fact, current Radom factory materials indicate that the stock was primarily intended to be used with a full-automatic version of the VIS, which was never produced due to Poland's adoption of the MORS submachine gun. The original drawings for the VIS shoulder-stock, with two different locking mechanisms, do exist and have been published in Poland. A few hundred reproductions of the better of the two were manufactured in Poland some years ago and marketed in both Poland and the United States. They are quite well made, and their locking mechanism works to tightly secure the stock to both widths of mounting groove. However, their legal status in the United States is uncertain due to the failure of the Bureau of Alcohol, Tobacco and Firearms (BATF, now BATFE) to issue a compliance letter for them. This holster stock opens at the back and has spaces for the pistol and for two spare magazines. It has a bar at the rear for the attachment of a belt loop.

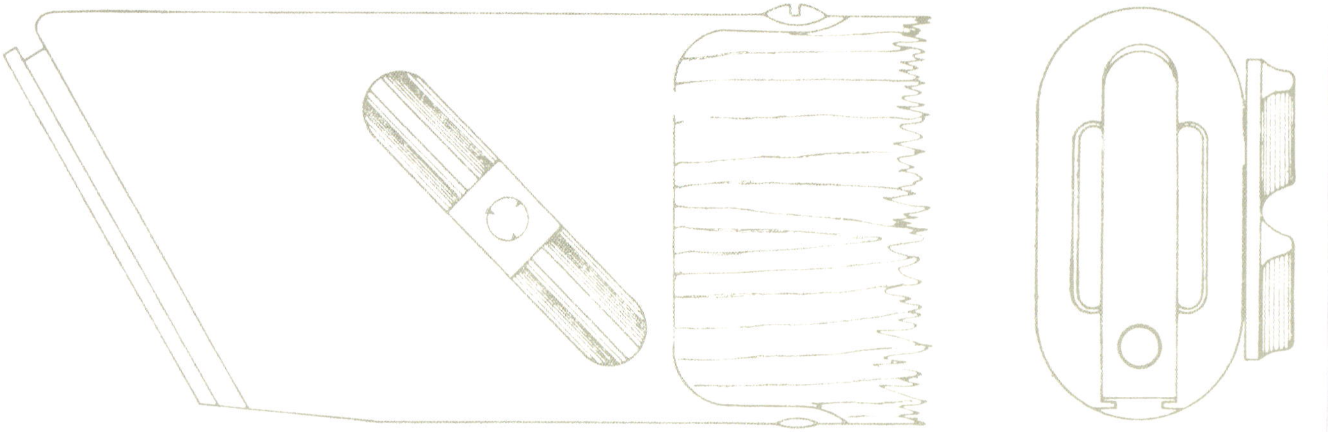

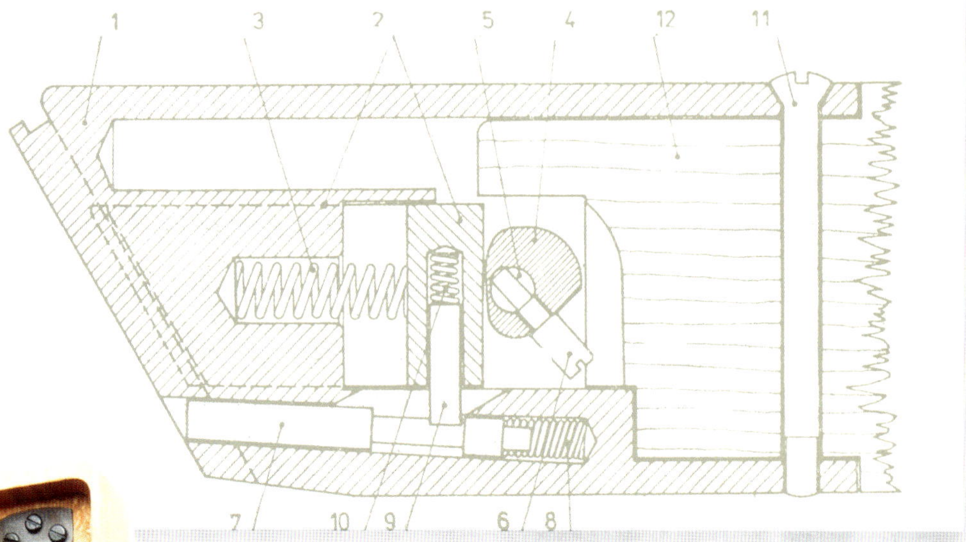

(Fig. 90-1) Drawings: Stock hardware variant with butterfly release button and expanding side ribs to eliminate play when the pistol is mounted. This system was overly complex.

(Fig. 90-2) Replica stock made according to the above plans: The shoulder-stock is hollowed out to accommodate the pistol and two magazines. (Photograph by Anthony Vanderlinden)

(Fig. 91-1) Replica holster-stock. No original VIS shoulder-stocks are known to exist, but the original plans for their construction surfaced in Poland some years back, and a quantity of exact replicas were produced and imported into the United States. As no BATFE letter has been issued legalizing these stocks, their attachment to a VIS pistol with a slotted mainspring housing can only be legally done if the combination has been registered as a short barreled rifle, and tax paid. (Photographs by A. Vanderlinden)

(Fig. 91-2) The butterfly locking system with expanding ribs shown both engaged and disengaged. (Photographs by A. Vanderlinden)

(Fig. 92-1) Drawings: Proposed hardware variant with simple release button. It was never manufactured.

(Fig. 92-2) Slab sided stock which fits the VIS: These surface from time to time and appear to be modified postwar Browning High Power stocks, or postwar manufacture in the High Power fashion. Fitting one to a slotted VIS will also create a short barreled rifle, albeit not an authentic appearing one. (Photographs by A. Vanderlinden)

Manuals

The prewar Polish manual covering the use and maintenance of the VIS has been reproduced and translated into English. Original copies are very rare. It is probable that a wartime German manual was issued, but the author has never seen a copy.

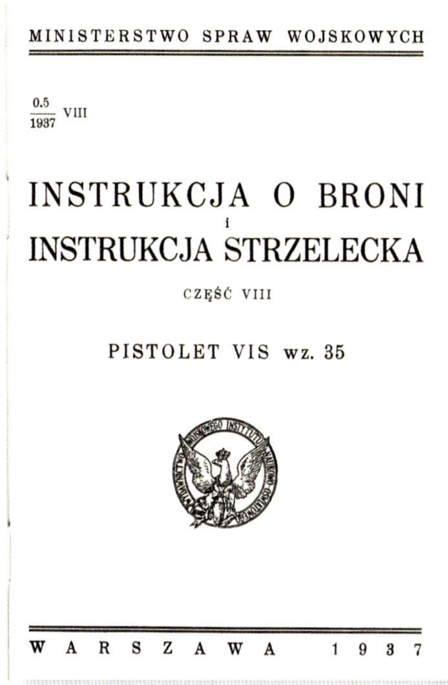
MINISTERSTWO SPRAW WOJSKOWYCH

0.5/1937 VIII

INSTRUKCJA O BRONI
i
INSTRUKCJA STRZELECKA

CZĘŚĆ VIII

PISTOLET VIS wz. 35

WARSZAWA 1937

Ammunition

Prewar Polish pistol ammunition is exceedingly rare, prewar 9mm boxes are even rarer. Wartime ammunition does surface on occasion, some being repackaged after the war.

(Fig. 93-1) A Polish produced 9mm Parabellum cartridge dated 1938. (Photograph by A.Vanderlinden, author's collection)

(Fig. 93-2) Left: A 1943 dated box of 16 9mm Parabellum rounds produced in German-occupied Poland (maker code kam). Right: A repacked box of German-occupied Polish produced 9mm Parabellum cartridges with a Finnish label. (Photographs by A.Vanderlinden, author's collection)

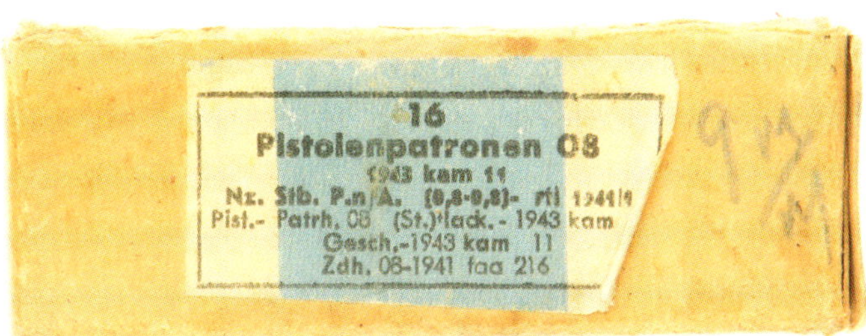

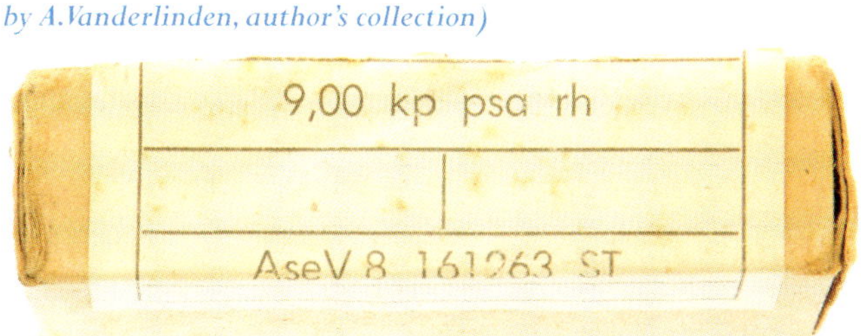

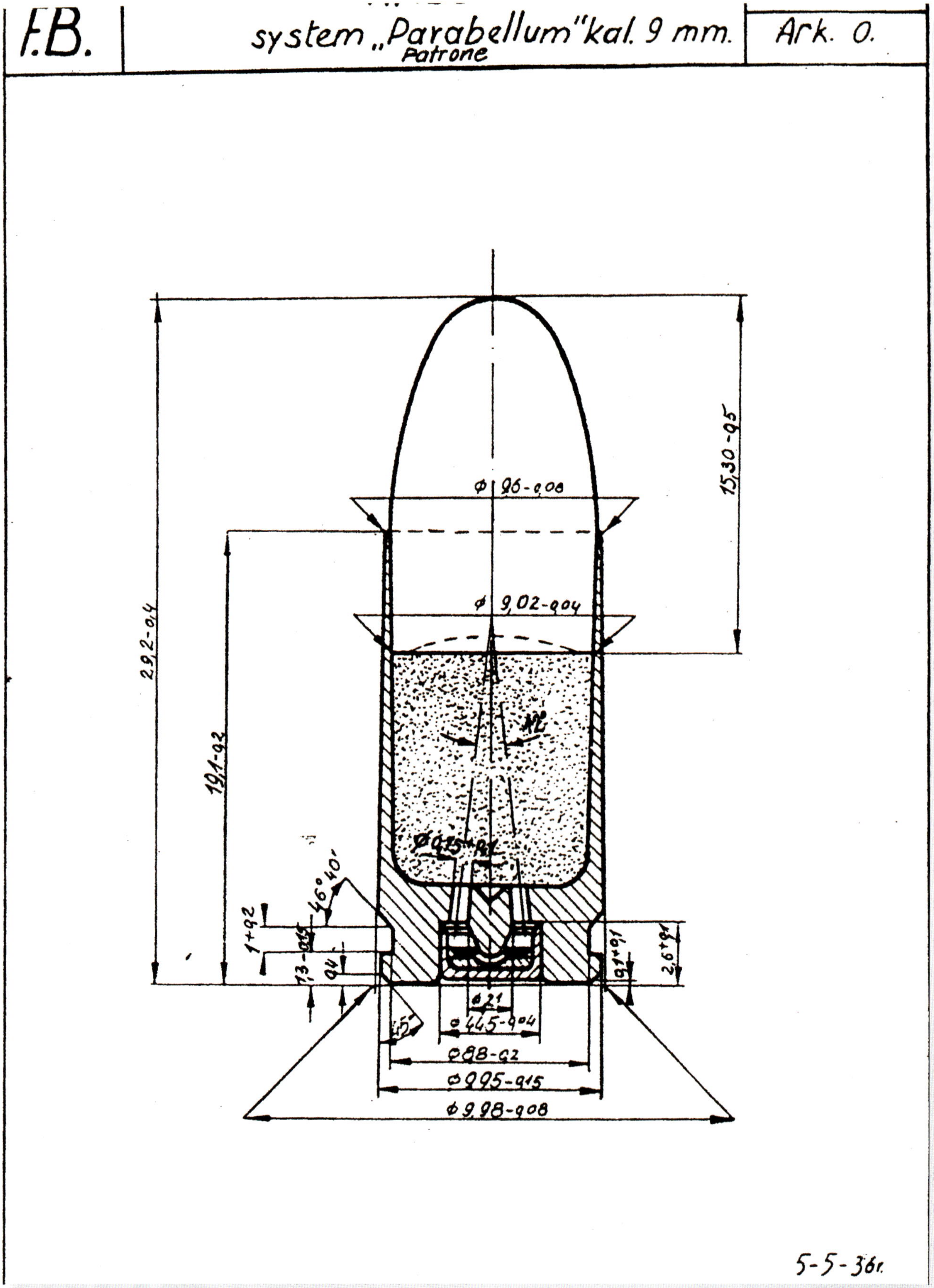

(Fig. 94-1) A May 5, 1936 dimensioned drawing of the 9mm Parabellum cartridge as produced in Poland.

.22 Conversion Unit

A .22 caliber conversion unit, consisting of a barrel insert and adaptors for the smaller cartridges was proposed in 1939 and is illustrated in select drawings. It was apparently never produced.

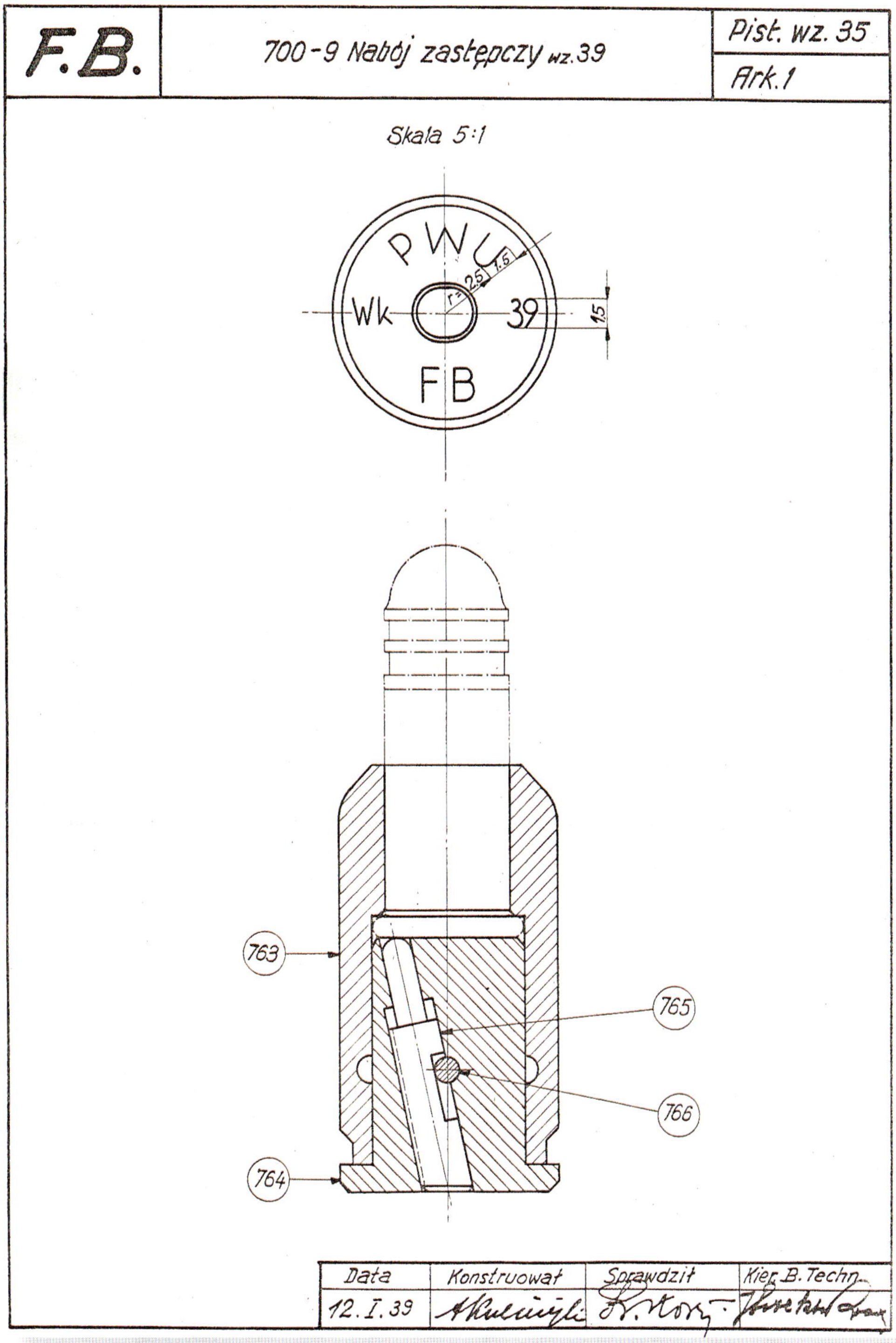

(Fig. 95-1) The cartridge holder for the proposed conversion kit, with internal rimfire firing pin.
(Page 96) Top: Detail of the conversion unit barrel insert, threaded at the front for a nut which would hold it in place.
Bottom: Detail of the conversion unit's firing pin, which converted a centerfire blow to rimfire.
(Page 97) Top: 20 meter target for the proposed VIS .22 conversion unit.
Bottom: Drawing of .22 conversion unit cleaning rod.
(Documents courtesy of Waclaw Ustupski)

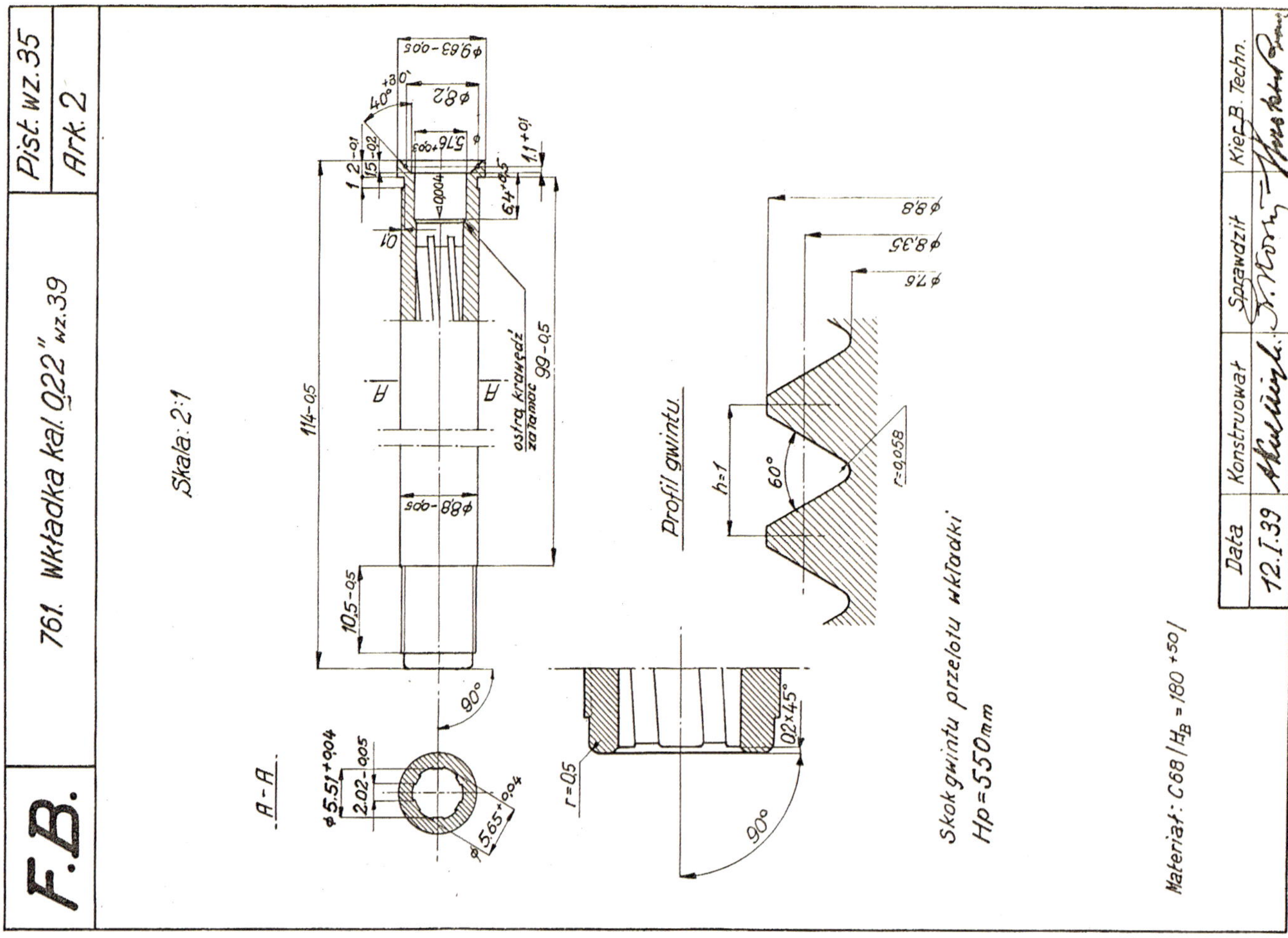

(Fig. 96) See fig.95-1

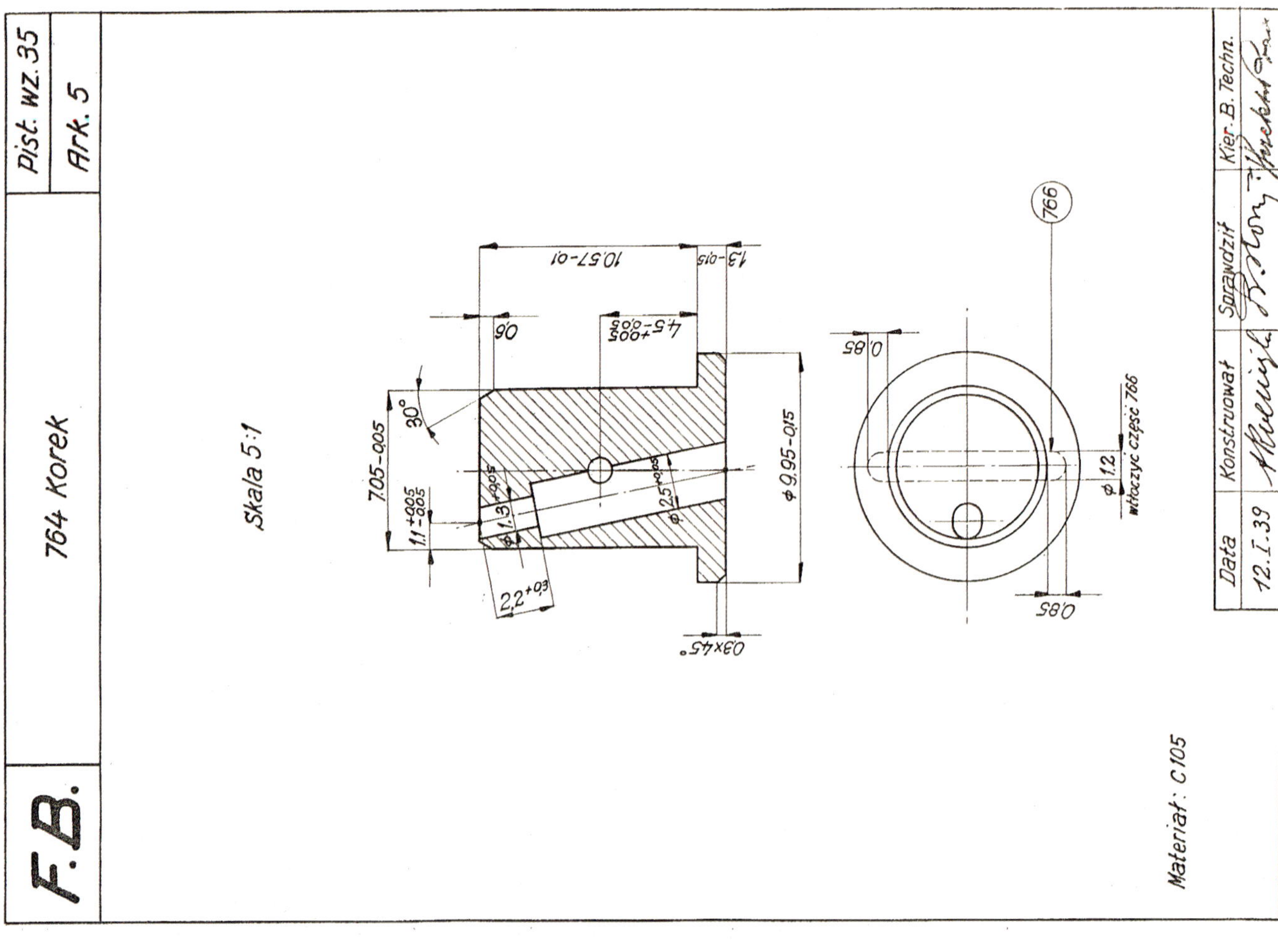

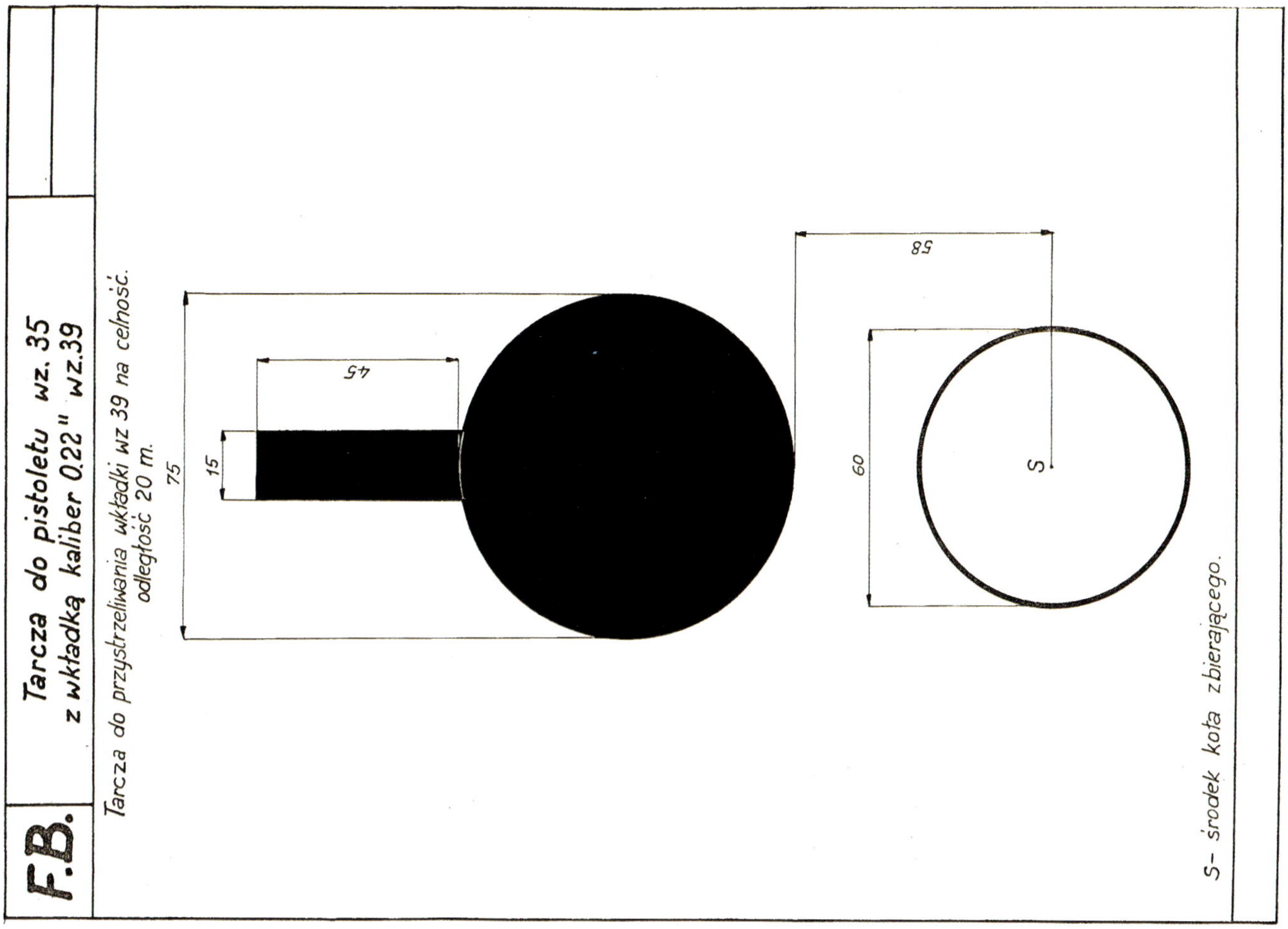

(Fig. 97) See fig.95-1

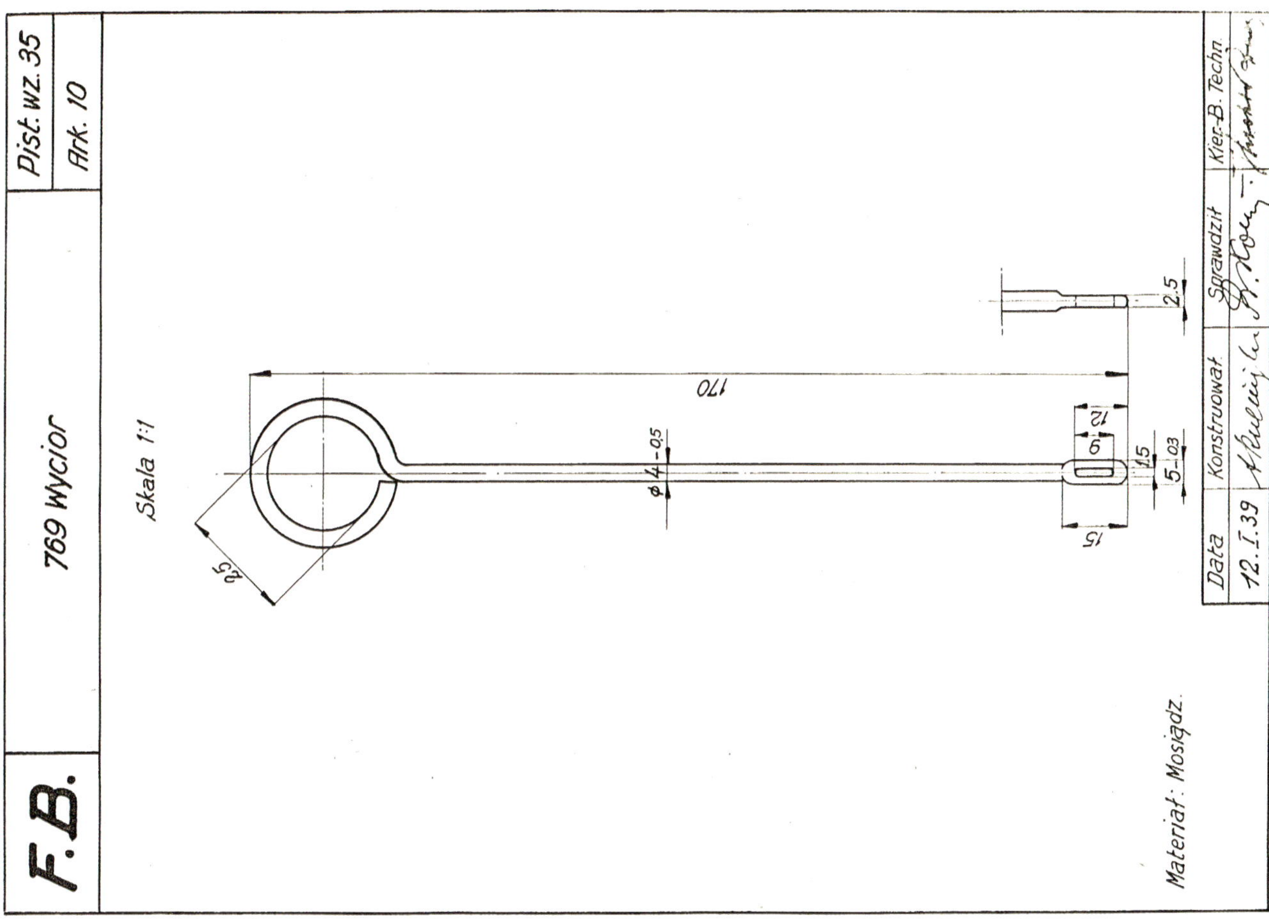

Chapter X
Shooting the VIS and a description of how it functions

Shooting the VIS

The VIS pistol has often been described as an extremely rugged arm, accurate, reliable, and of strong design and construction. We agree. One most interesting such discussion is an article by Jim Thompson on pp.17-21 of the 1987 Gun Digest. Mr Thompson clearly favors the shooting qualities of the pre-War VIS, but praises the shootability of all VIS pistols. It is the opinion of the present author that owners and collectors should not shy away from shooting these pistols if in good condition and proper ammunition is used- so long as a given gun's condition as a collector's item is not affected. I have done it, and can assure the reader that the VIS is a fun pistol to shoot.

If a decision to shoot a VIS in one's collection is made, a competent gunsmith should do a prior safety check of the weapon. An interesting discussion in this vein may also be seen in the publication TROUBLESHOOTING YOUR HANDGUN, which calls the complicated captive recoil spring the only weak point of the VIS.

How the VIS functions

To load the pistol, a loaded magazine is inserted into its well in the grip until it is held fast by the magazine catch. The slide is then pulled back to the limit of its travel and released. The recoil spring now powers the slide forward, stripping the top cartridge from the magazine and forcing it into the chamber at the rear of the barrel. The pistol is now held firmly and the trigger pulled. As the grip safety is depressed, the trigger is able to push against the bottom flat of the disconnector. If the slide is fully forward, the top of the disconnector is in its recess on the bottom of the slide, and the trigger's motion is conferred to the sear. The sear now swings on its pin and releases the hammer which, powered by the mainspring, swings forward and strikes the firing pin. The firing pin is driven forward, and strikes the primer, firing the chambered cartridge.

As the fired bullet moves forward, recoil energy causes the slide and barrel, which are locked together, to move to the rear, compressing the recoil spring at the same time. When the sloped camming surface at the rear of the barrel lug encounters its mating surface in the frame, the rear end of the barrel is cammed downward. This disengages the locking surfaces at the top of the barrel from the slide, and once the frame's cam stops the barrel's rearward motion the slide continues rearward to the limit of its travel. As the slide travels to the rear, the extractor withdraws the newly fired cartridge case from its chamber in the now stationary barrel, and once the slide passes over the ejector, the base of the fired case strikes the ejector and the empty cartridge case is thrown clear of the pistol.

The compressed recoil spring now drives the slide forward, stripping another cartridge from the magazine. The cam on the front of the barrel lug now cams the barrel forward into locked position with the slide, and the slide's disconnector recess is once more positioned for firing. Now the pistol may be fired once the trigger is released, which action recycles the disconnector into position to move the sear and allow the hammer to drop once the trigger is once again pulled.

This cycle may be repeated until the magazine is empty, which will cause the slide to be held open by its latch, alerting the shooter that the pistol is empty. Now the magazine catch may be depressed, the empty magazine may be replaced with a loaded one, and the slide latch depressed, to allow the slide to run forward and chamber another cartridge, repeating the process.

Chapter XI
Take-down of the VIS

Field-Stripping

NOTE: BEFORE HANDLING ANY FIREARM, YOU MUST BE CERTAIN THAT THE WEAPON IS UNLOADED! SPECIFICALLY, ON THE VIS THE MAGAZINE MUST BE REMOVED AND THE SLIDE WITHDRAWN AND LATCHED IN ITS REARWARD POSITION WITH THE SLIDE LATCH. THE CHAMBER END OF THE BARREL SHOULD THEN BE VISUALLY INSPECTED TO ENSURE THE ABSENCE OF A CARTRIDGE. THIS MUST BE DONE EVERY TIME A FIRE-ARM IS HANDLED.

Beginning with the slide held to the rear by the slide latch; on earlier pistols the slide should be released and held in position so the takedown lever may be rotated into its notch on the slide. On later wartime pistols wherein the hammer drop lever acts as a takedown lever, the hammer drop lever is depressed and the slide is allowed to slowly run forward until it is held by the internal hammer drop lever bearing against the notch on the hammer's upper surface.

In either case, a semicircular notch in the slide's lower left rail will now be aligned with the pin of the slide latch. To remove the slide latch (remember, the magazine should be out of the pistol) the pistol is held with its left side downward. Now, the protruding end of the recoil spring guide rod is grasped and pulled out a short distance. The slide latch will often drop out of the pistol at this point. If it does not, a nudge with a finger should free it. Note that the slide latch is held in place by the back piece of the captive recoil spring bearing against a flat surface on its pin. An attempt to pry the latch free without releasing the recoil spring's hold may break the latch-slide latches with their thumb pieces broken off are not uncommon, and result from improper takedown. Replacement slide latches are hard to find.

On earlier pistols, the slide and frame should now be held while the takedown lever is rotated downward, and the slide slid forward off of the frame. On later wartime pistols, the slide is pulled to the rear, releasing the hammer drop lever from the hammer notch, and the slide is likewise slid forward off of the frame. The slide is now held or placed upside down, and the rear piece of the captive recoil spring rotated ¼ turn in either direction. The captive spring may now be removed to the rear as a unit, then the barrel may be removed to the rear from the slide.

Field stripping is now complete.

Reassembly is done in reverse order. Reinstallation of the slide latch is done with the slide held in place by the takedown or hammer drop lever. The pistol is positioned right side down, and the slide latch held on its pin hole while the spring guide is pulled as before. The latch should fall into place, requiring only minor positioning with a finger.

Takedown 1

Takedown 2

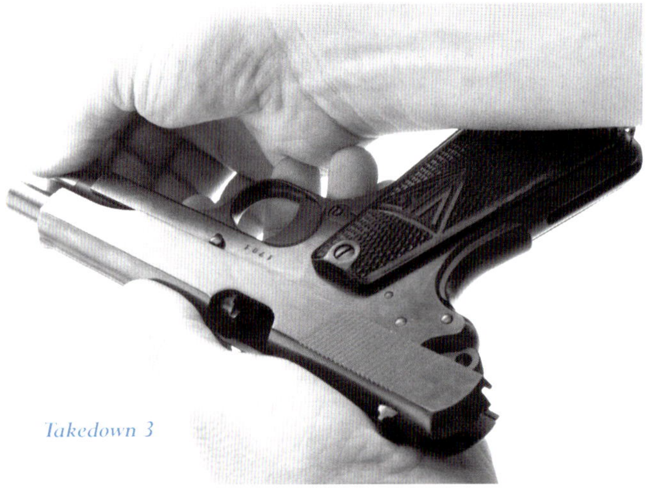

Takedown 3

Takedown 4

Takedown 5

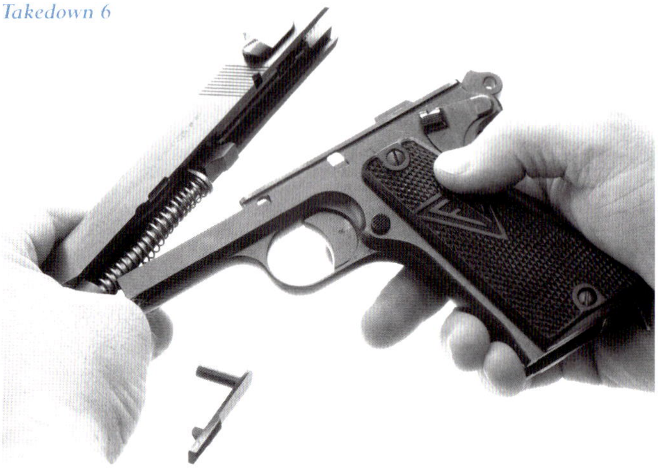

Takedown 6

Takedown 7

Takedown 8

Takedown 9

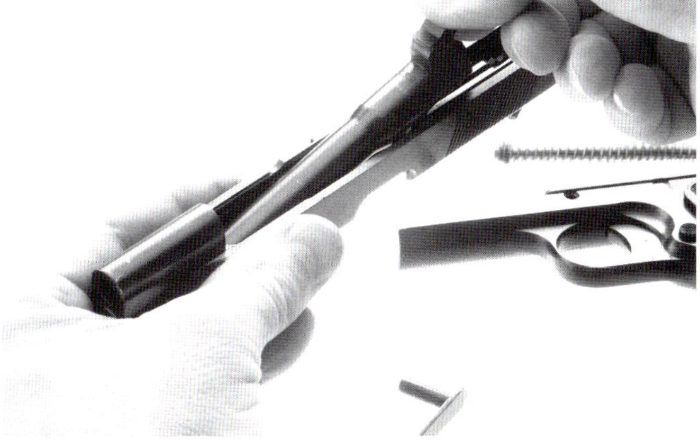

Reassembly 1

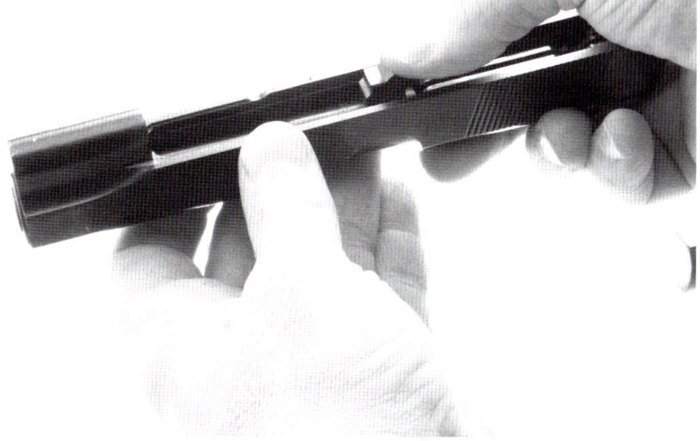

Reassembly 2

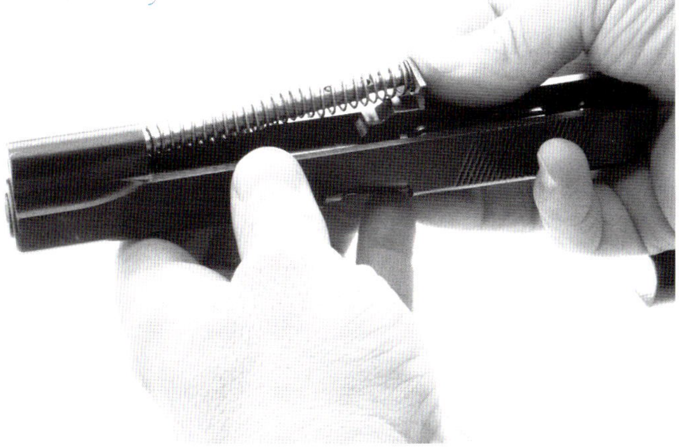

Reassembly 3

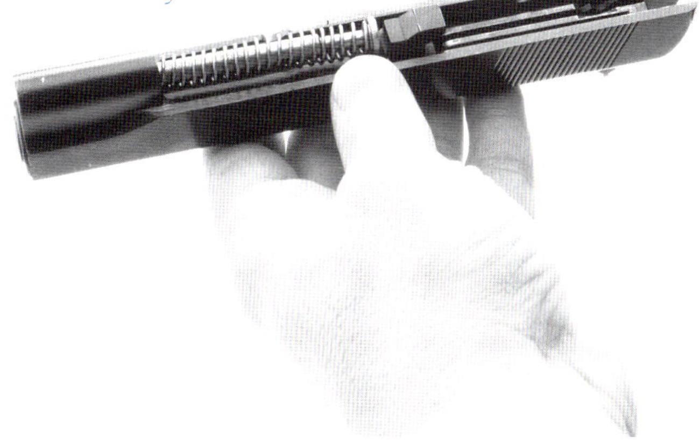

Reassembly 4

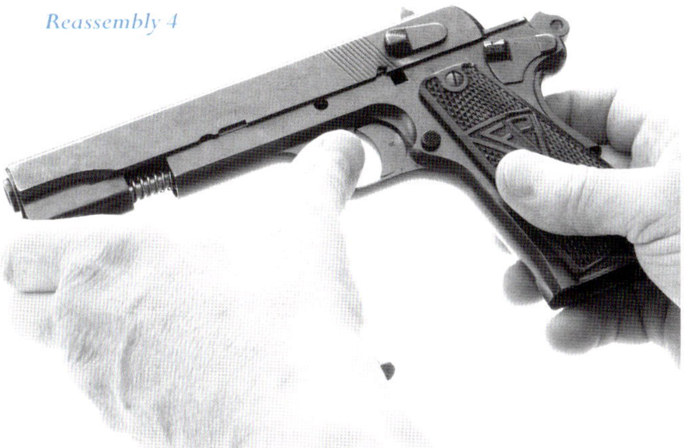

Reassembly 5

Reassembly 6

Reassembly 7

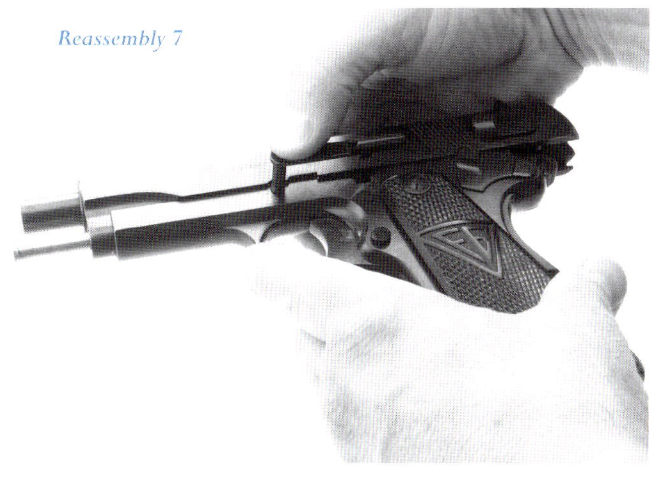

Reassembly 8

Reassembly 9

Detail Stripping

The Frame Assembly

The VIS frame assembly rather closely resembles that of the Colt M1911 pistol. The grips on the VIS must, however, be removed in detail stripping. This is accomplished by removing the four grip screws. On earlier pistols with grip screw bushings, the bushings are ordinarily staked in place and should not be removed unless replacement is called for.

With the grips removed and the hammer down, the mainspring housing pin may be removed in either direction- it is not held tightly unless rust or coagulated lubricant are a factor. The mainspring housing may now be removed by sliding it down off the frame. This housing may be further disassembled by drifting out the small lateral pin at its upper side. This frees the mainspring and its cap, but this should only be done if replacement of the mainspring is necessary, or if the pistol has suffered water or similar damage. Once the mainspring housing is removed, the combination sear-disconnector, trigger, and grip safety spring may be lifted out of the frame. On earlier pistols the takedown lever should now be rotated to a vertical position. This allows a small lug on its pin to line up with a notch on the pin's hole in the frame, and the lever may now be pulled out of the frame. As the takedown lever pin also retains the grip safety, the grip safety may now be removed to the rear.

The hammer pin may now be removed from right to left, and the hammer removed along with its strut and pin. It is usually not necessary to drift out the small pin and separate the hammer from its strut. The sear and disconnector pin is also removed from right to left, and sear as well as disconnector should fall free if the frame is tilted front end upward. Sometimes a jiggle on the trigger is required to free these parts.

The VIS trigger is retained in the frame by the magazine catch. To remove the magazine catch, it is pushed in by its thumb-piece and its lock (a small part on the right side with a screwdriver slot) is rotated counterclockwise until the catch is freed. This may be done with a small screwdriver or with the bent tip of the combination spring, it sometimes takes some doing to get the lock into its slot on the side of the catch. Once the catch is removed, the trigger may be removed to the rear. It will often simply drop free.
The magazine catch lock and its spring are separated from the catch by rotating the lock clockwise until it is freed. It is under spring pressure and should be retained during this process.

The VIS ejector is a small, flat steel part held in place on the frame's upper surface, above the disconnector hole, by a thin lateral pin. It may be removed by drifting out its pin, but should not be removed unless it is damaged. Its removal is not easy, as the pin is often frozen in place.

In the case of later pistols, which lack the takedown lever and have riveted pins retaining the grip safety, hammer, and sear-disconnector, detail stripping of these parts should ordinarily be avoided unless necessary, as to replace broken parts. It should be kept in mind that drifting out the riveted pins will slightly expand the pin holes in the receiver. This may be minimized by drifting the pins in the same direction, and only far enough to free the parts they secure. Afterward, they may be returned to their places and re-staked. The mainspring housing and magazine catch (with one exception illustrated in our catalog section, see N° K8228 on pages 223-224) are the same in late and in early guns with the sole difference being the lack of a stock attachment slot on all but early wartime pistols.

When reassembling the frame assembly, the relationship of sear and disconnector should be kept in mind. Also, care should be taken not to cross-thread the grip screws, especially the later large screws which install directly into the frame without a bushing. The grip screws should also be kept in order, as they are individually filed to length to avoid interference with the magazine. If trouble is encountered with a magazine being hard to insert and withdraw, and the magazine itself is in good condition, the problem may often be traced to a too-long grip screw or screws.

The Slide and Captive Recoil Spring

The VIS slide assembly consists of the slide body, rear sight, extractor, firing pin and spring, hammer drop lever and spring, and firing pin retainer (or stop).

The rear sight is held in a lateral dovetail and may be removed or adjusted by drifting with a non-marring tool. This should be kept to a minimum to avoid loosening the sight in its slot.

The other listed parts are all held in place by the firing pin stop. Unlike that in the Colt M1911 pistol, the stop in the VIS is somewhat different in form, and is installed quite tightly ("dead tight") in the slide. It is tight for a reason- so the hammer drop will positively not cause the firing pin to be struck by the falling hammer when the hammer drop lever is actuated. To free the retained parts, the firing pin stop must be moved downward while the firing pin is being depressed. The firing pin's rear portion extends through the hole in the stop when the pin is at rest, and would prevent the stop from being moved downward. The best way that the author has found to remove the stop is by using a U.S. Krag tool, which has a lateral pin of just the right length and diameter to enter the firing pin stop hole and push the firing pin out of the way, while the tool may be tapped downward to move the stop. Once the stop is drifted about 1/8 of an inch downward, the hammer drop lever must be removed before the firing pin stop can be removed any further. This is an easy task, only requiring a little jiggling to free the lever return spring from its place in the slide. Now the firing pin stop may be drifted the rest of the way down and removed.

(Fig. 103-1) Use of U.S. Krag tool in slide takedown

The firing pin and its spring can now be easily removed. To remove the extractor, it should be started out from the front by use of a non-marring tool or drift. Once started on its way, the extractor's retaining notch may be used to lever it out from the rear of the slide.

This whole procedure should be approached with caution, especially when dealing with late war pistols. The firing pin stop in these late weapons may be brittle, and may crack on removal. Admittedly, detail stripping of the slide is often necessary to replace a broken firing pin or extractor.

One other point which must be reemphasized at this time is that the VIS firing pin stop is supposed to be very, even "dead" tight. This is a safety measure in that it prevents the firing pin from moving when the hammer drop feature is used. If the stop is loose, it is possible that the pistol has been "messed with" , and a competent gunsmith should be consulted before firing the pistol. Firing pin length is also critical to the proper functioning of the hammer drop mechanism, and this should also be checked by a gunsmith, especially if it is suspected that the firing pin has been replaced.

The captive recoil spring assembly has already been described above, and it is recommended that it not be disassembled. As has been stated, this assembly is held together by a small, dumbbell-shaped pin which rides in a slot in the back piece rod or "tail". This small pin may become cocked inside the hollow guide rod during an attempt at disassembly, and may gall or otherwise damage the inside of the guide rod, or the slotted rod in which it rides. This may ruin the captive recoil spring assembly, which is a hard to find part. Further, the hollow guide rod may be found staked on both sides, to the rear of the lateral assembly holes for the dumbbell shaped pin. This acts to keep the "tail" of the back plate properly aligned, but also prevents the dumbbell-shaped assembly pin from being removed unless force is used to push the internal rod, the "tail" through the staked area. In other words, this assembly is meant to be permanent, especially where staking is present.

If, once the above is considered, it is deemed absolutely necessary, the captive recoil spring assembly may be disassembled as follows. The assembly should be compressed lengthwise in a non-marring clamp or vise, to the point where the retaining pin is laterally loose in its hole. This indicates that it is in line with the expanded assembly point in the internal rod's slot. Reaching this point may take some force, and must be done carefully. Once the internal retaining pin is visible on both sides of the spring rod, the holes on both sides should be examined to see whether one is larger. This is because some of the internal dumbbell shaped retaining pins appear to be tapered. If this is the case, removal and reassembly must be from the larger hole. In any event, some manipulation will likely be necessary to ensure that the internal retaining pin is properly lined up for removal. Once this is done, the recoil spring proper should be slid out of the way (this requires some strength) and the retaining pin may be drifted out with a small punch. The pin is quite small and easily lost, so it must be put in a secure place. Now the vise or clamp may be relaxed and the captive recoil spring assembly will come apart.

Assembly of the captive recoil spring is in reverse order, but it is not easy! When replacing the dumbell-shaped pin, vise grips or a clamp may be used to hold the recoil spring out of the way on its guide rod. This does not make reassembly easy, but it helps. See the following illustrations.

Reassembly of the slide is also in reverse order. When replacing the firing pin, the flat surface on its side must be lined up to interact with the hammer drop. The point at which the hammer drop lever and its spring are to be replaced is clearly visible through the lever's mounting hole, and assembly should be straightforward.

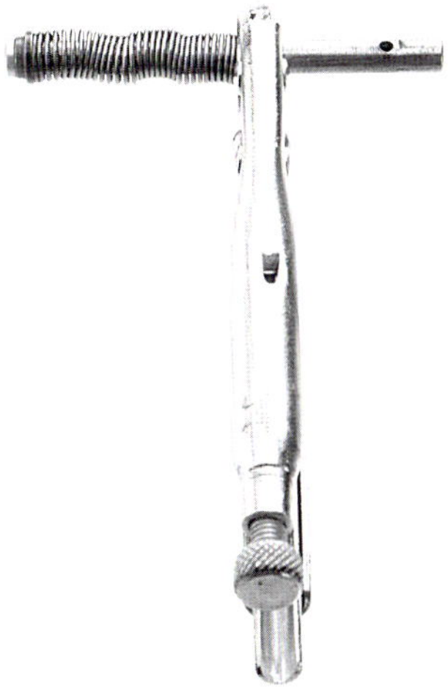
(1)

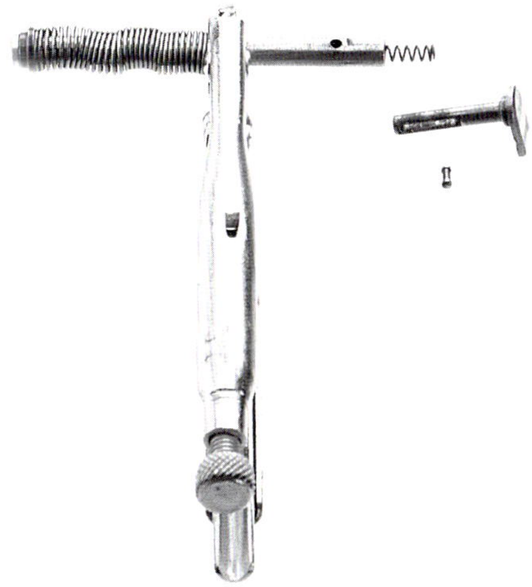
(2)

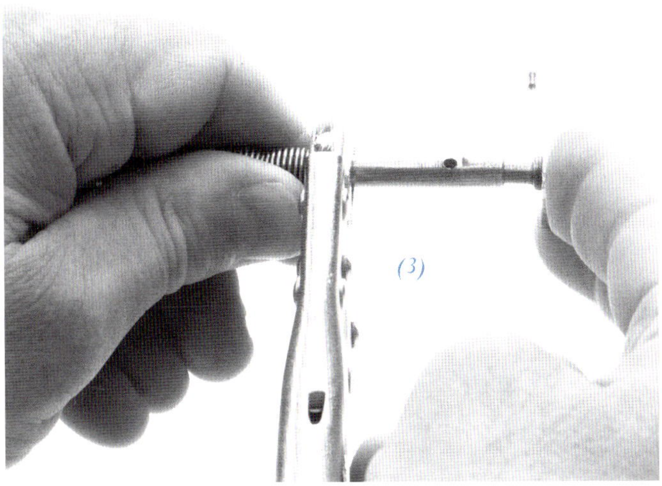
(3)

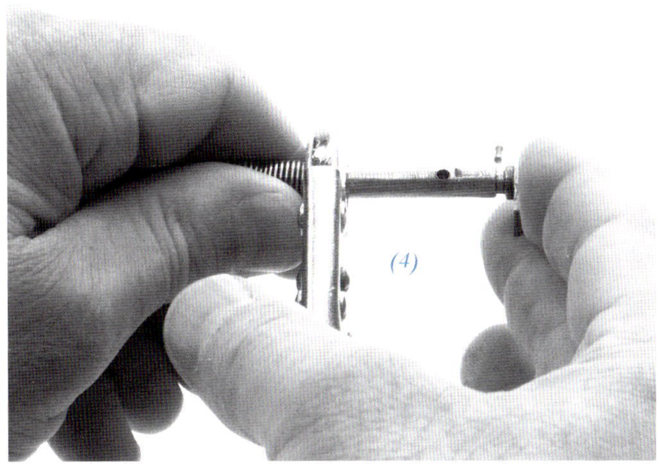
(4)

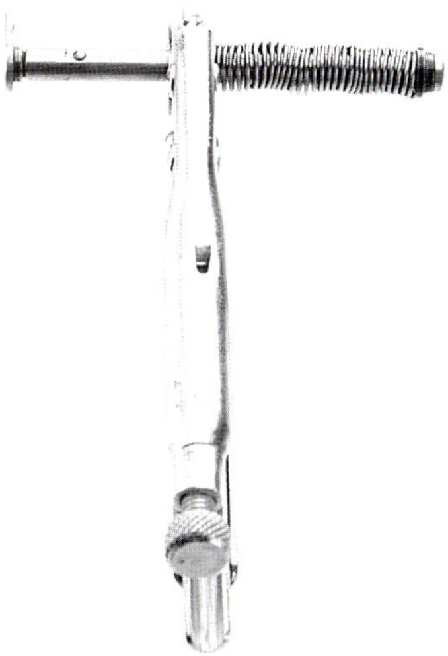
(5)

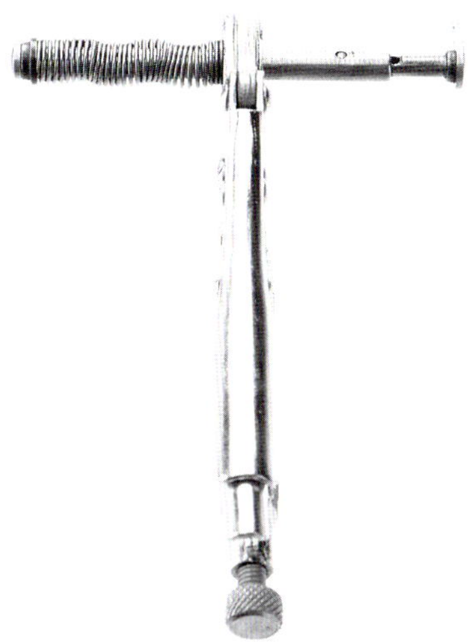
(6)

Parts Compatibility between the VIS and the Colt M1911

On page 21 of the 1987 GUN DIGEST, Mr. Jim Thompson in his excellent article on shooting the VIS, states that "…[s]ome odds and ends of Colt parts will fit. Grip screws and bushings and mainsprings, for example, fit directly, and some pins and springs can be adapted without undue work…"

This inspired the author of this book to detail strip a VIS and a Colt M1911 side-by-side, dial caliper in hand. The results follow.

1.Grip screws and bushings have the same thread on both pistols. VIS screws and bushings are a little smaller externally, but the Colt screws and bushings can easily be modified to be exact replacements for the VIS.

2.Firing pin springs are identical. Mainsprings and mainspring caps differ, but the Colt spring can be shortened and the cap modified for satisfactory replacement in the VIS.

3.The Colt hammer pin is the same diameter as that on the VIS, and is the same diameter as the VIS mainspring housing pin. The Colt pin can be shortened to serve as a VIS hammer pin, and if its head is removed it can serve as a VIS mainspring housing pin.

4.The Colt magazine catch lock has the same head diameter and pin diameter as that on the VIS, and the Colt and VIS magazine catch springs are also of the same diameter. The Colt's pin and spring are longer, as the Colt's frame is wider, but both Colt components can be shortened and used in the VIS.

While Colt hammers are similar and of the same thickness, the full and half cock notches on Colt and VIS hammers are differently positioned, and the hammers do not appear to be compatible. VIS rear sights, however, are not easily replaced by Colt 1911 parts- the VIS sight dovetail is narrower than the Colt.

The above shows just how closely the VIS was influenced by the Colt, especially in light of the fact that the grip screw and bushing threads are non-standard U.S. types. Of course, these observations will also be helpful to U.S. owners who need to repair their pistols.

This is also a good place to make mention of the fact that certain VIS parts have been reproduced. Mr. Thompson of the 1987 Gun Digest article had barrels reproduced in three grades. The author has one of these in the most utilitarian grade, and can attest to its quality and usability. It is not believed that these barrels are still on the market, but they may still be occasionally encountered. VIS extractors and firing pins have also been reproduced, and are available. Both should be fitted to the individual pistol, and if properly installed should work well, The reproduction firing pins the author has seen are made somewhat differently from the originals, but they are internal parts and any differences would not be externally visible, especially on a "shooter grade" VIS. There are also replacement grips on the market. Some are excellent replicas, some are not! New magazines are also available, but should be checked for fit and function. These magazines are easy to spot, as they generally have stamped followers quite unlike those on earlier production.

It should be kept in mind that the VIS was, in its major components, a product of hand fitting, and that therefore the interchangeability of barrels, slides, grip safeties, and mainspring housings is not assured. Hand fitting is often required, and should be done by a competent gunsmith who knows this pistol. The hammer drop lever must be individually fitted to ensure that it will properly withdraw the firing pin to a safe position before the hammer is released, Likewise, replacement firing pins need to be fitted for proper length and a proper relationship to the hammer drop cam. Minor parts, to include the slide latch, grip screws and bushings, pins and springs will usually drop in. Extractors usually require some fitting. Modern replacement extractors are made too long at the rear, and must be shortened and contoured to match the individual slide. If a firing pin retainer needs to be replaced, it MUST be individually fitted very tightly to the slide, as improper fitting of these parts can cause an unsafe condition, as stated above.

(Fig. 107-1) VIS (L) and M1911 Colt (R) frames compared.

(Fig. 107-2) Colt magazine catch parts above VIS magazine catch parts.

(Fig. 107-3) VIS mainspring housing pin (L), and Colt hammer pin (R).

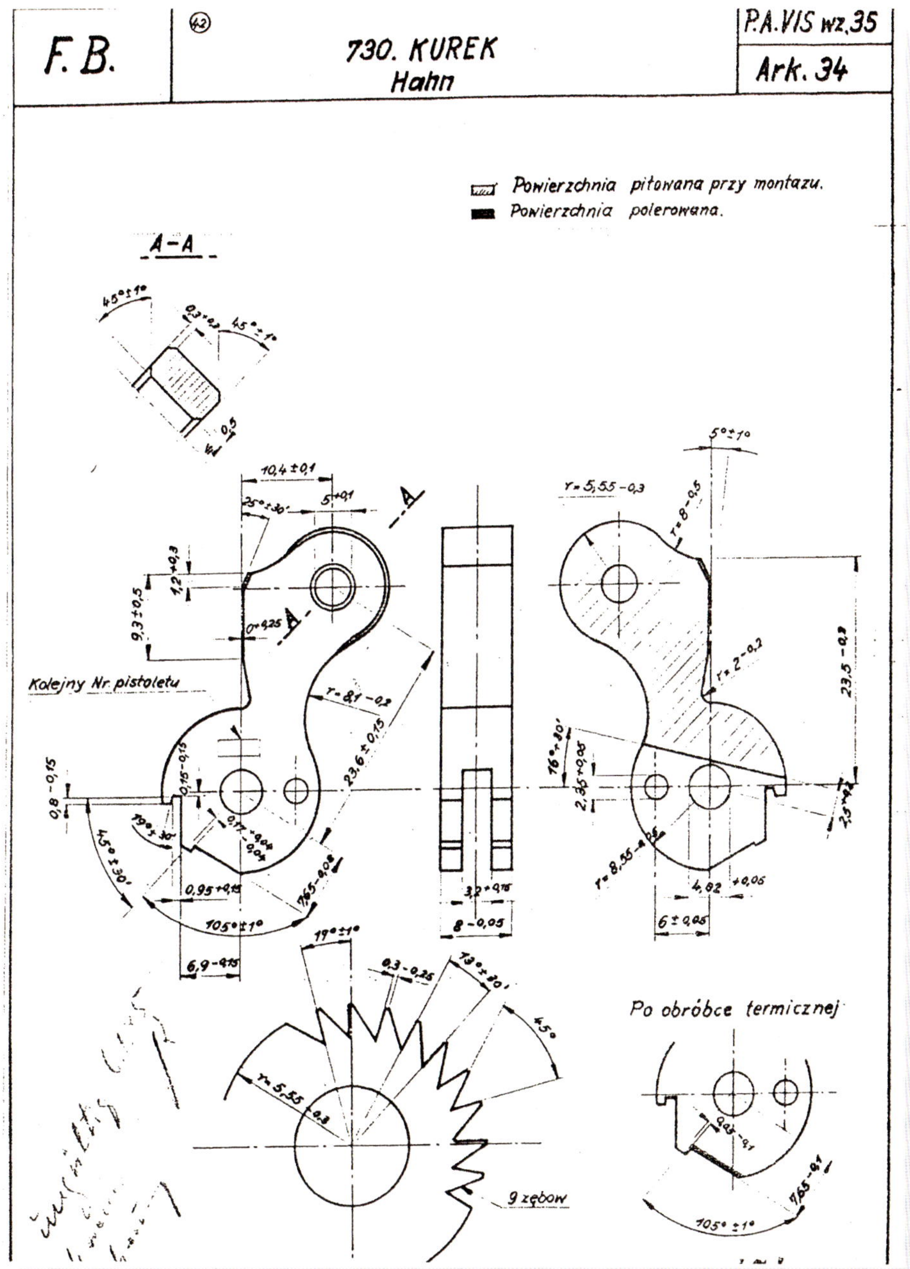

(Fig. 107-4) Factory drawing of VIS hammer

Chapter XII
Collecting the VIS

What makes the VIS an interesting, even a fascinating collector's item is the variety in which it exists. While it remained basically the same weapon throughout production, it exists in two basic forms: Polish and German. The pistol's Polish form may be divided by year of production, and the advanced, or lucky, collector may find the rare prototype or presentation pistol. Then, even the not so advanced collector can ponder the odd features of very late Polish pistols, which were produced at a time when the invading Germans were literally at (or beyond) the gates of the Radom factory.

Collectors of the German version of the VIS begin their quest with weapons produced just after the German seizure of the Radom plant. Basically, these are Polish pistols with German markings added. Beyond these rare examples, there exist a wealth of variations: German Navy marked pistols, early and late production examples, "lunch box" pistols made from stolen parts by the Polish underground, U.S.G.I. captured pistols with stories, East German marked pistols. The list goes on to include 1945 Steyr production, "BNZ" marked pistols, pistols devoid of slide markings, and other "oddballs". The advanced collector may also be interested in acquiring one pistol with each serial number letter prefix. This is decidedly not an easy task. Nonetheless, the collector has considerable help when it comes to being aware of VIS variations.

It has been stated that the wartime VIS is the one pistol which most clearly illustrates the gradual decline of German industry as the Second World War went from victory to defeat. This is certainly true, and this progressive decline has led students of the VIS to divide its production into time of production- dependent groupings of weapons sharing similar characteristics. Such categorization can be of positive benefit to collectors. However, it should be bourne in mind that many of these weapons were produced under adverse, and sometimes desperate, wartime conditions. Under such circumstances it was reasonable for the factory to use all available parts in the assembly process, so if some parts from a "prior" variation were found, a later numbered pistol would be made incorporating them. In late production, the next part on top of a parts pile or bin would be used, without regard to the niceties of "later categories".

One should also keep in mind that VIS deliveries were not necessarily made in serial number sequence and in some instances "later" numbers may have been produced before "earlier" ones. This was in fact the practice at U.S. weapons plants when under wartime conditions. In the author's view this is an additional unknown which makes a study of the VIS pistol all the more interesting.

VIS categories have been put forward by authors in the United States and in Poland. Major Robert D. Whittington III and Robert J. Berger have divided German VIS production into three grades, with Berger adding Steyr production as a fourth. Jan Still goes into more detail, dividing three grades of the German made pistols into eight subvariations, with serial number ranges given for each. The Polish authors Zbigniew Gwozdz and Piotr Zarzycki, simply divide German production into early, mid, and late war. As may be seen, all agree that there are three general types of German produced VIS pistols to be found, marked by the presence or absence of progressive production shortcuts. The categories described above will be elaborated upon by illustration herein, with some of our own comments to follow.

(Fig. 108-1) Mr. Emil Walega, a veteran of the Polish Resistance, in his gun room in Poland. (Photograph courtesy of W. Ustupski)

(Fig. 109-1) Part of Waclaw Ustupski's award winning VIS display at an Ohio Gun Collector Association display meeting. Right: Waclaw Ustupski in prewar Polish uniform at his OGCA meeting display. (Photographs courtesy of W. Ustupski)

Import Marked Pistols

VIS pistols recently imported into the United States will be found to be marked on the frame with the importer's name and location as well as other information. These markings are required to be visible, but may be in an unobtrusive location such as the frame bottom by the magazine opening. As of the date of this writing, pistols with these import marks are accorded a lower collector value than unmarked specimens, although this may not be the case if a given piece is especially rare. Whether these marks will so affect value in the future, or whether they may become a sort of sub-collectible in themselves, is anyone's guess.

A Warning on Counterfeits

Any industrial product can be duplicated, and if it has sufficient value, chances are that it will be duplicated by some unscrupulous character. A prime example is the "BNZ" pistol. The author has seen a correct "BNZ" slide fraudulently renumbered to match a late frame and barrel. This is easy to do on late pistols because the slide number is on the bottom of the left rail and not internal, as in earlier pistols. Caveat Emptor!

Reproduction Polish holsters have also been fraudulently "aged" and passed off as original. This would apply to German holsters as well, and especially to Navy or SS markings on any VIS holster. Experts should be consulted before money is lost. Guarantees from reputable dealers help a lot.

German Navy marked pistols can easily be faked by a skilled engraver, and should be approached with caution. It should be remembered that most Navy used pistols were not marked. To the point, Still in his work AXIS PISTOLS lists only 25 known Navy marked VIS pistols. Expert advice should be sought, one rule of thumb is that if a particular rare pistol has a provenance dating back to before it was especially valuable, it is more likely real. Yet, this is no guarantee!

Chapter XIII
Production Evolution through Observations & References to the Photo Gallery

The following lists observations of when changes were implemented and how these relate to the photography gallery. It is essential to emphasize that these are only guides especially during the occupation production. Many wartime variants exist and features can easily be observed outside of accepted/observed serial ranges, this can all be attributed to the "parts bin" approach of assembly where old parts were assembled together with new parts causing a significant overlap in features. Collectors should be reminded that wartime manufacturing does not fit into tightly controlled categories of specific features. Manufacturing and assembly of the VIS was chaotic; relying on various plants and assembly locations while manufacturing, assembly and transportation of parts and pistols were subjected to Allied air raids.

Prewar Polish Manufacturing Changes

Carbon steel replaces alloy steel frames: Alloy steel frames were assembled onto the first 3000 pistols. Carbon steel frames replaced alloy steel frames early in 1936 production. The difference in metal composition is not visible. (See pages 117-118)

Round sight replaced by a ramp-style front sight: The round front sight was only used during the first year of production (1936). A ramp-style front sight replaced the round sight in early 1937. (See pages 117-120)

Mismatched pistols assembled during the German invasion: An estimated 2000 mismatched pistols were completed in assembly. Serial numbers were not sequential. Many of these pistols lack a lanyard ring. (See p. 138-141)

German Occupation: Manufacturing changes and production simplifications - First Alphabet Prefix Series.

Elimination of the shoulder-stock slot: First observed on pistol N° B3318 but note that there is significant overlap as earlier produced frames were used up. (See pages 158-160)

Red Bakelite grips introduction: Earliest observed C6739, red Bakelite grips are scattered throughout production until replaced by the late war Steyr grooved grips. (See pages 159, 173, 194, 195, 198, 203)

Change from WaA77 to Eagle 77 markings: Observed on early "L" serial number prefix block with significant overlap into the "N" serial number prefix block. (See pages 176-180)

Finger clearance cuts eliminated from frame manufacturing: First reported in the mid "J" serial number prefix block, but first noted by author in the mid- "K" serial number prefix block. (See page 174)

Elimination of fine polishing of metal before bluing: First observed in the "G" serial number prefix block but with significant overlap as many finished parts kits remained available. (See page 169)

Elimination of the early "P35(p)" stamping near the slide legend: First observed early in the "K" serial number prefix block but with significant overlap. (See pages 176-178)

Introduction of Ersatz pressed wooden grips: First observed in the mid- "R" serial number prefix block. As the grips were made from low quality (soft) wood, many were most likely replaced, making it difficult to establish where these were introduced in the production timeline. (See page 185)

Grasping grooves on the left side of slide is increased from 10 to 11: Observed starting in the mid- "R" serial number prefix block. (See page 186)

Elimination of milling the slide rib behind the rear sight: First reported is pistol N°T2410 without slide rib. (See fig.111-2 and pistol N°T6371 on page 188)

Introduction of the fabricated trigger: Observed early in the "W" serial number prefix range, with overlap. (P. 191)

Transition from small to large grip screws: First reported in the later "W" serial number prefix block but with considerable overlap into the "Z" serial number prefix block. (See pages 193 and 194)

Elimination of the takedown lever: Observed early in the "Z" serial number prefix block with overlaps in the entire "Z" serial number prefix block and into the second alphabet prefix "A" serial number prefix block.(Pages 192-197)

Introduction of inventory numbers on the frame and slide of parts kits: First reported in the late "T" serial number prefix block but with significant overlap, unnumbered frames or slides continued to appear into the "U" serial number prefix block until the earlier unnumbered parts were used up. (See fig. 66-1)

German Occupation: Manufacturing changes and production simplifications - Second Alphabet Prefix Series.

Introduction of grooved wooden grips: First observed on pistol N° A7110. The grooved grips coincide with the introduction of the phosphate finish. These first show up randomly in the "A", "B" and "H" prefix, yet it should be noted that these were most likely assembled at the time of the "H" letter code. (See page 199, 210, 211)

Steyr/Znaim parkerized finish introduction: Observed N° A7110. These first show up randomly in the "A", "B" and "H" serial number prefix block, yet it should be noted that these pistols were most likely assembled at the time of the "H" serial number prefix block (as some do no longer have E/77 markings - see page 211). Both blue as well as phosphate pistols appear in the mentioned prefix letter codes.

Introduction of WaA623 instead of E/623 as final acceptance marking next to firing proof on slide: Observed in the late "B" serial number prefix block. (See page 201)

Steyr made frames and slides with WaA623 or E/623 markings (instead of E/77) markings: First appear in the "J" serial number prefix block. The WaA623 or E/623 marking is stamped on the trigger guard. These specific markings were indicative of which plants or subplants made the parts. Assuming that these slides were made at Steyr/Gusen and not Znaim. (See page 212)

Replacement of WaA623 with E/623 as final acceptance on slide next to firing proof: This is believed to indicate a change in where the parts were sourced, most notably the slide. It is possible that this indicates the beginning of Znaim production from entire new (raw) materials, while previously Znaim finished and assembled kits were leftover from the Radom produced inventory. Some overlap can be expected. (See page 217)

Location of slide serial number moved from under the slide to the top of the slide rail: This change was introduced with Steyr/Znaim slide production and is found on slides made entirely from new/raw materials at Znaim in the "K" serial number prefix block. (See fig. 70-2)

Introduction of E/623 marked magazines with skeleton/stamped followers: Assumed to be in the "K" serial number prefix block. Depending on remaining parts inventory, some magazines may have been assembled into phosphate "A", "B" and/or "H" serial number prefix pistols.

Solid recoil spring assembly introduction: It is generally accepted that these appear throughout the "K" serial number prefix block, although overlap into earlier letter codes is likely, depending on availability of earlier articulated recoil spring assemblies.

"bnz" marked pistols: This variant appears in the early "K" serial number prefix block. (See page 70 and 214 for additional details)

Elimination of firing proof and final acceptance marking: Found on last ditch pistols. (see pages 225-226)

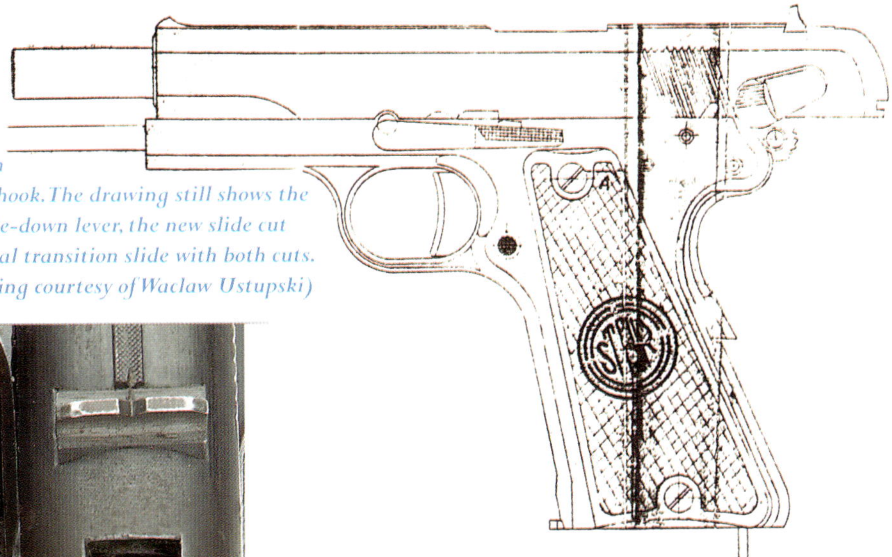

(Fig. 111-1) A Steyr drawing illustrating the function of takedown by hammer-drop lever and hammer hook. The drawing still shows the early slide cut for operation with take-down lever, the new slide cut was added. See fig. 197-1 for an actual transition slide with both cuts. Note the Steyr marked grips. (Drawing courtesy of Waclaw Ustupski)

(Fig. 111-2) Comparison between the early milled slide rib (right) and late slide rib (left). (Photograph by Anthony Vanderlinden)

Photography Gallery I:

The Polish Production from Prototype through 1939

(Fig. 112-1) A Free Polish soldier with his VIS fighting along British troops. Many Polish soldiers fled together with the Polish Airforce after the capitulation of their home land, first to France and eventually to England. The prewar VIS pistols that made it to England remained in service with Free Polish troops during the war. (Photograph courtesy of Waclaw Ustupski)

Polish Production: Prototype VIS N° 0063

(Fig. 113-1)
Prototype Number: ***0063***
Final Inspection Proof: ***None***
Notes: ***VIS prototype pistol N° 0063 has remained in Poland through the German Occupation and Communist eras, to the present day. It was hidden by being buried, possibly in a barn. Many firearms were so hidden in Poland, thus this pistol shows considerable corrosion. This is common to such resurfaced firearms in Poland. This should be considered honorable corrosion. Note that the front end of this pistol's recoil spring is even with the front end of its slide. Compare to pages 114-115.***
Photographs courtesy of Waclaw Ustupski, from the Emil Walega collection

Polish Production: Prototype VIS N° 0108

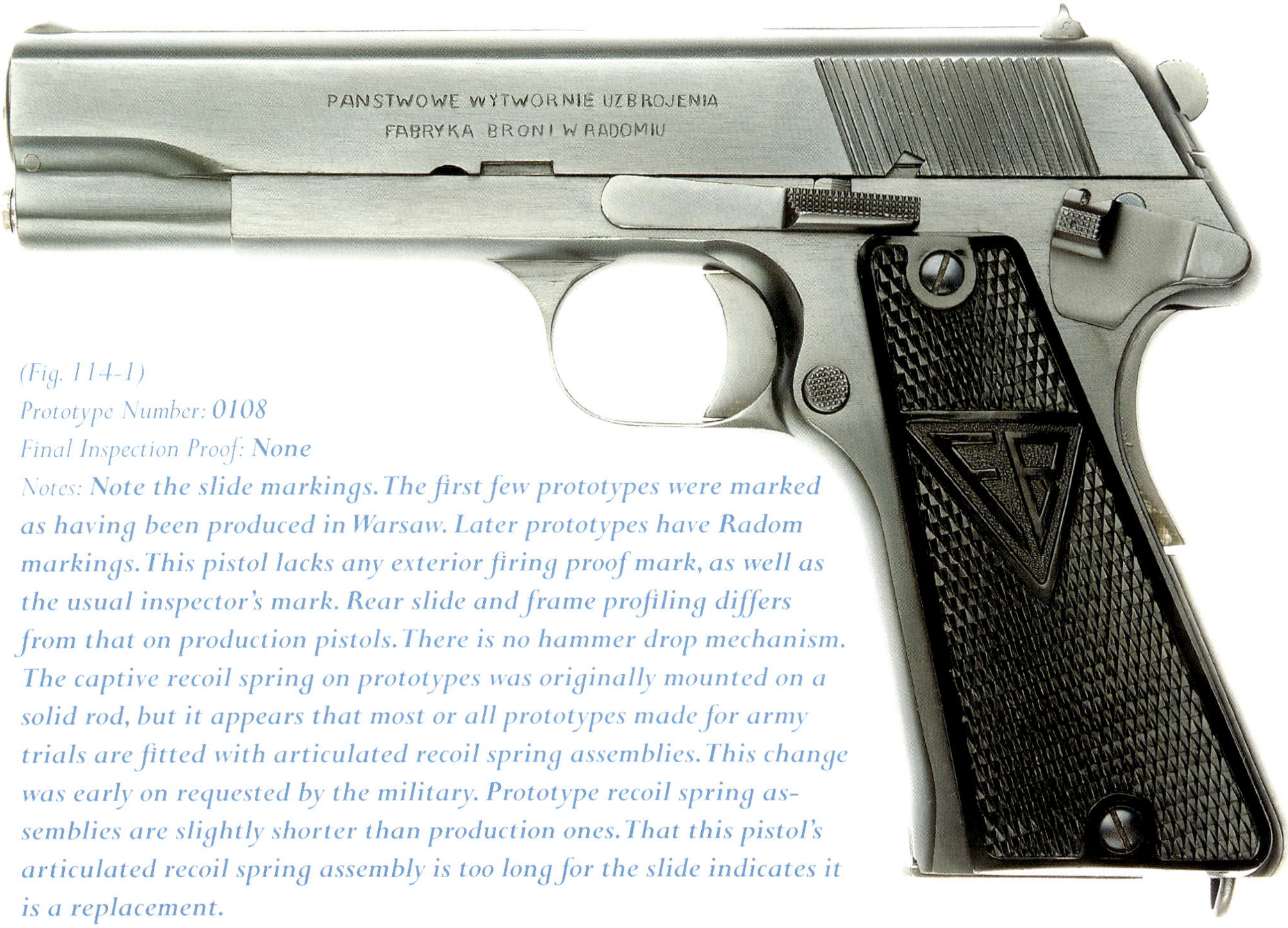

(Fig. 114-1)
Prototype Number: ***0108***
Final Inspection Proof: ***None***
Notes: ***Note the slide markings. The first few prototypes were marked as having been produced in Warsaw. Later prototypes have Radom markings. This pistol lacks any exterior firing proof mark, as well as the usual inspector's mark. Rear slide and frame profiling differs from that on production pistols. There is no hammer drop mechanism. The captive recoil spring on prototypes was originally mounted on a solid rod, but it appears that most or all prototypes made for army trials are fitted with articulated recoil spring assemblies. This change was early on requested by the military. Prototype recoil spring assemblies are slightly shorter than production ones. That this pistol's articulated recoil spring assembly is too long for the slide indicates it is a replacement.***

Notes continued: ***The pistol's sights, hammer, ejector, firing pin, firing pin stop, sear and disconnector also differ from those on production VIS pistols. This pistol is one of a reported 134 prototypes produced. Prototype VIS pistols showed remarkable accuracy and durability, firing 6000 shots without any breakage, as well as retaining a high degree of accuracy after 5000 shots fired.***
Photographs by A. Vanderlinden, courtesy Waclaw Ustupski collection

(Fig. 115-1) Notes: ***Translation- "State Weapon Factory Weapon Factory in Radom"***

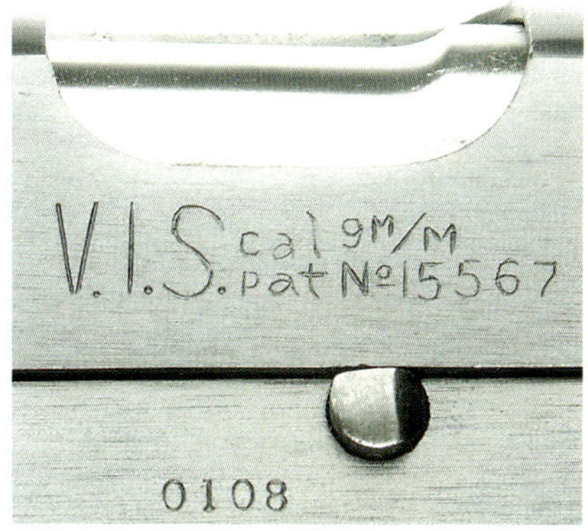

(Fig. 115-2) Notes: ***This shows the pin which secures the barrel bushing, as well as the protruding front of the captive recoil spring's guide rod. Prototype recoil spring assemblies were slightly shorter than those on production pistols, due to a longer barrel lug on prototypes. This pistol has a replaced recoil spring assembly.***

(Fig. 115-3) Notes: ***This shows the rowel style hammer without hole on the prototype VIS, as well as a side view of the ejector.***

(Fig. 115-4) Notes: ***A top view of the rear frame, showing workmen's marks as well as the configuration and location of the ejector. It is quite similar to the Colt M1911 ejector. The prototype sear and disconnector also closely resemble those on the Colt M1911.***

(Fig. 115-5) Notes: ***The prototype's serial number as well as the only external inspector's marking on the front right triggerguard are shown here.***

Polish Production: Prototype VIS N° 108 (continued)

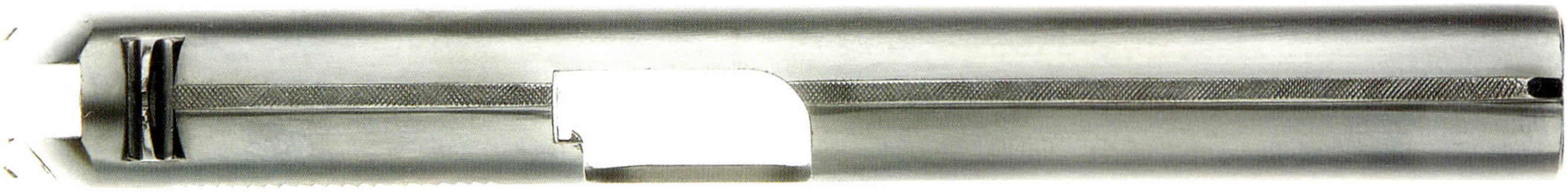

(Fig. 116-1) Notes: *The front sight and top rib are signifigantly wider than those on production pistols. The front sight also has a side profile sloping toward the pistol's rear, unlike the round profile on 1936 VIS production. The firing pin stop can also be seen to be of Colt M1911 type. The firing pin itself is also of Colt type, and lacks the flat surface added to its side to allow interaction with the hammer drop lever on production pistols. The bottom of the slide may be seen to lack any internal markings other than the serial number. Note the square rear sight opening. The checkering on the top of the slide is different from later production models.*

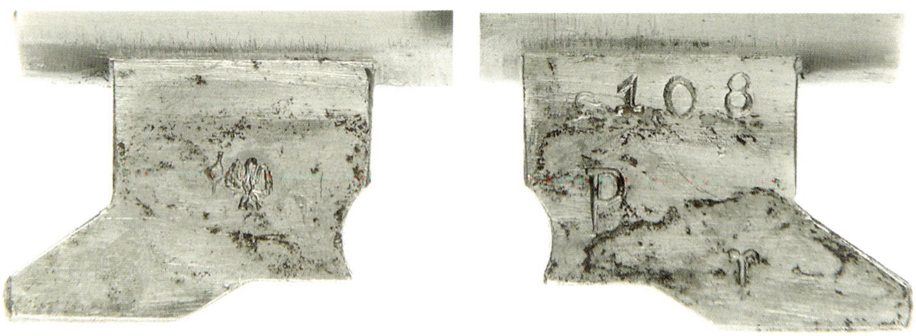

(Fig. 116-2) Notes: *Markings on the prototype barrel lug include a serial number, Polish Eagle (probably a firing proof), and inspector's mark. The barrel's underlug is longer and more massive than that on the production VIS (compare e.g. fig. 25-1). It is this increased length which has caused the replacement, later recoil spring guide rod to be slightly too long for the prototype slide. It also causes the takedown cut on the slide to be slightly forward of that on production pistols- compare fig. 114-1, upper photograph, with fig. 117-1, upper photograph.*

(Fig. 116-3)
Notes: *The prototype magazine is almost identical to production magzines with only dimensional differences.*

Polish Production: 1936 Eagle VIS N° 0702

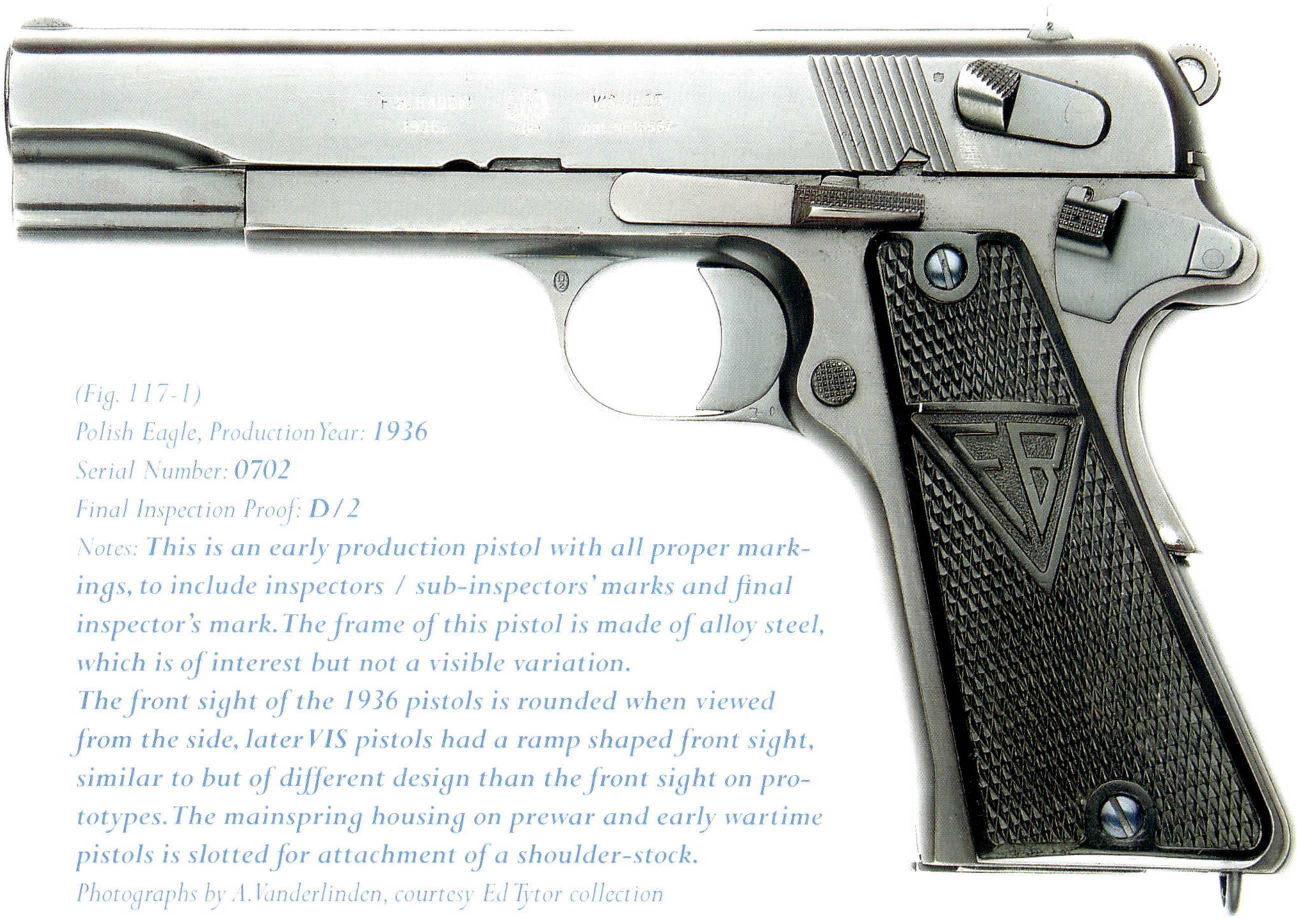

(Fig. 117-1)
Polish Eagle, Production Year: ***1936***
Serial Number: ***0702***
Final Inspection Proof: ***D/2***
Notes: ***This is an early production pistol with all proper markings, to include inspectors / sub-inspectors' marks and final inspector's mark. The frame of this pistol is made of alloy steel, which is of interest but not a visible variation.***
The front sight of the 1936 pistols is rounded when viewed from the side, later VIS pistols had a ramp shaped front sight, similar to but of different design than the front sight on prototypes. The mainspring housing on prewar and early wartime pistols is slotted for attachment of a shoulder-stock.
Photographs by A. Vanderlinden, courtesy Ed Tytor collection

Polish Production: 1936 Eagle VIS N° 1440

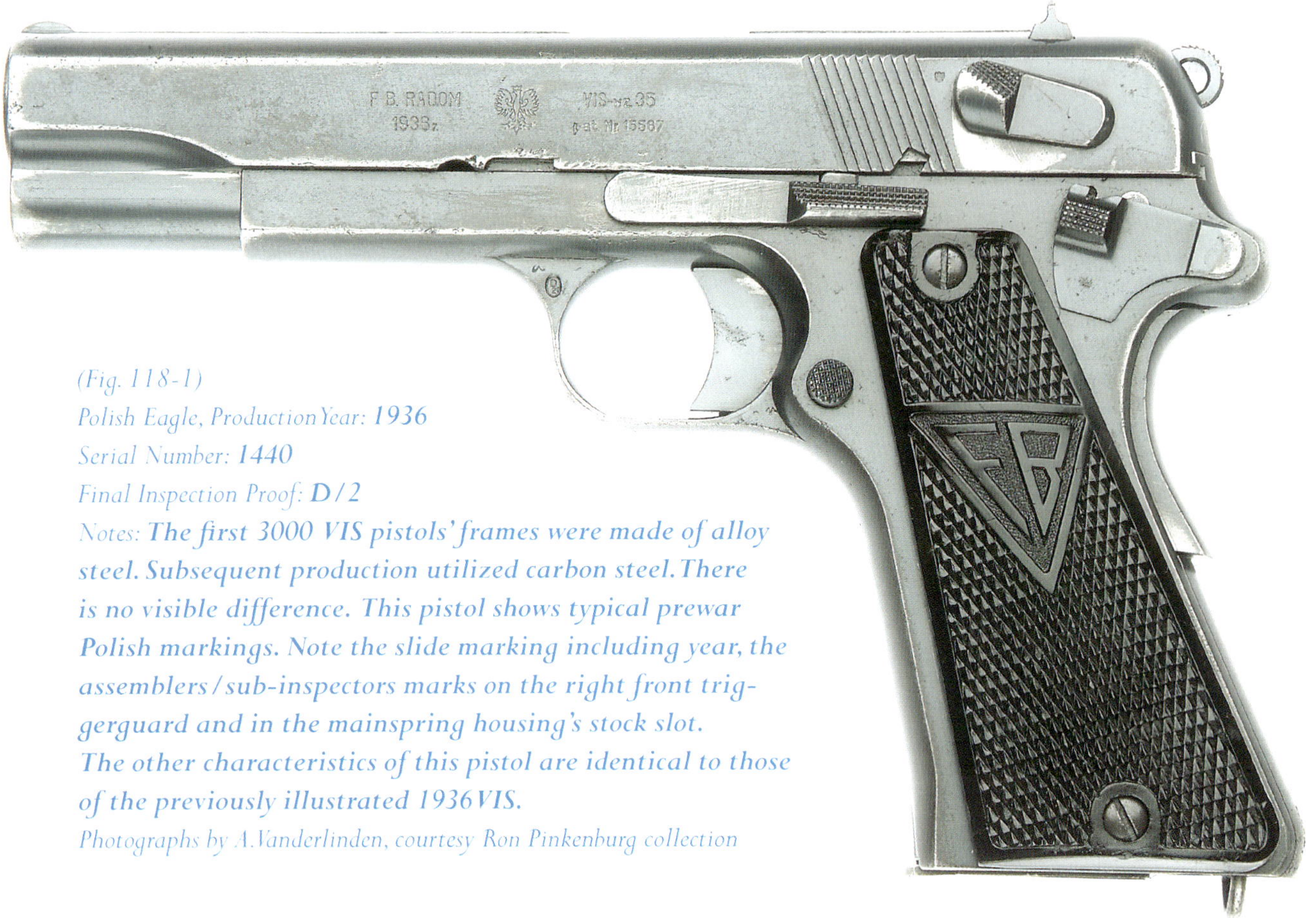

(Fig. 118-1)
Polish Eagle, Production Year: ***1936***
Serial Number: ***1440***
Final Inspection Proof: ***D/2***
Notes: ***The first 3000 VIS pistols' frames were made of alloy steel. Subsequent production utilized carbon steel. There is no visible difference. This pistol shows typical prewar Polish markings. Note the slide marking including year, the assemblers/sub-inspectors marks on the right front triggerguard and in the mainspring housing's stock slot. The other characteristics of this pistol are identical to those of the previously illustrated 1936 VIS.***
Photographs by A. Vanderlinden, courtesy Ron Pinkenburg collection

Polish Production: 1936 Eagle VIS N° 3314

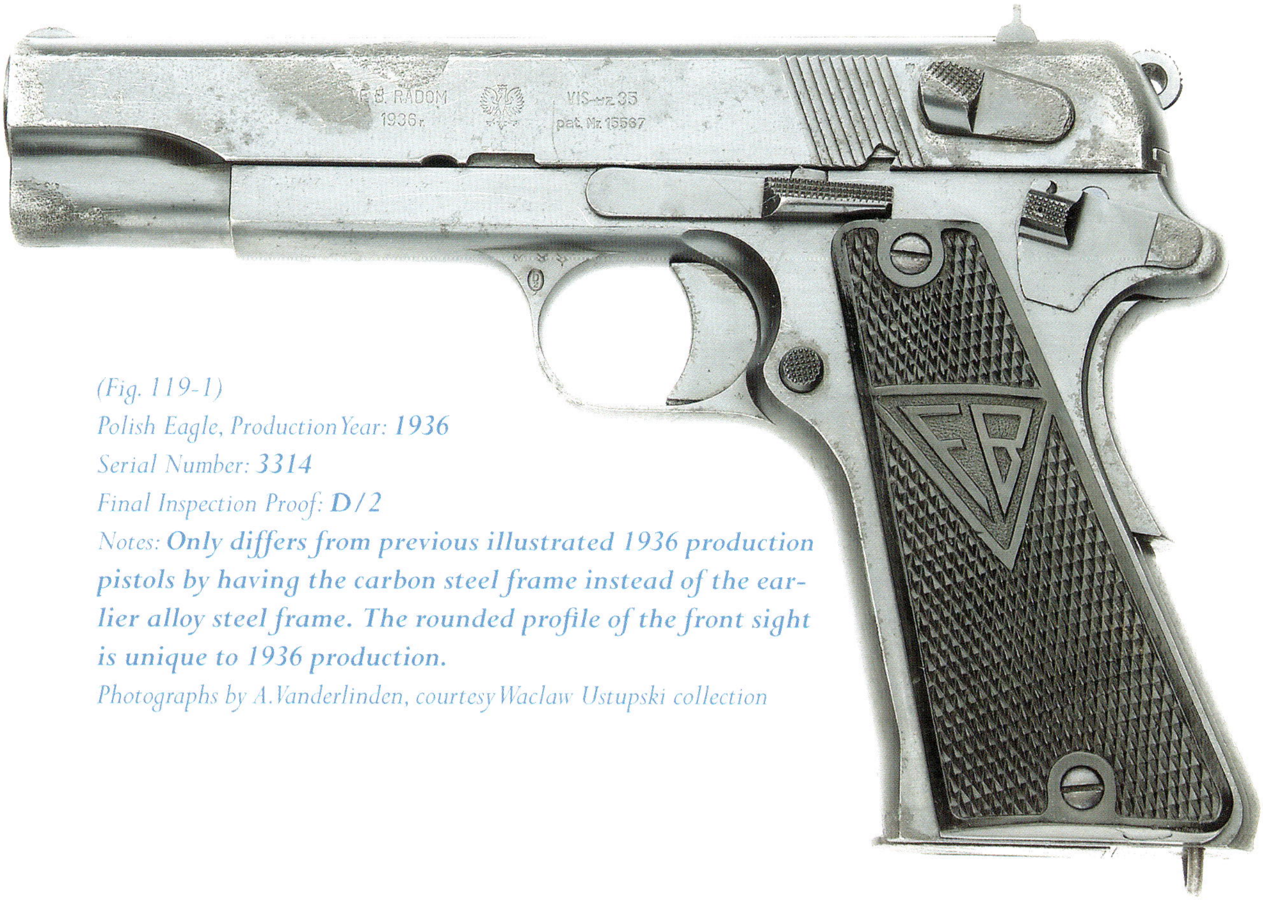

(Fig. 119-1)
Polish Eagle, Production Year: ***1936***
Serial Number: ***3314***
Final Inspection Proof: ***D/2***
Notes: ***Only differs from previous illustrated 1936 production pistols by having the carbon steel frame instead of the earlier alloy steel frame. The rounded profile of the front sight is unique to 1936 production.***
Photographs by A.Vanderlinden, courtesy Waclaw Ustupski collection

Polish Production: 1936 Eagle VIS N° 4256

(Fig. 120-1)
Polish Eagle, Production Year: ***1936***
Serial Number: ***4256***
Final Inspection Proof: ***D/2***
Notes: ***The rounded profile of the front sight is unique to 1936 production. This pistol shows a number of workmen's marks on the frame and trigger. The right grip has been broken and imperfectly glued together.***
Photographs by A. Vanderlinden, courtesy Waclaw Ustupski collection

Polish Production: 1937 Eagle VIS N° 5359

(Fig. 121-1)
Polish Eagle, Production Year: ***1937***
Serial Number: ***5359***
Final Inspection Proof: ***D/2***
Notes: ***This pistol shows refinishing as well as British proof marks; indicative of import or export to or from the United Kingdom. A prior owner has added a large ring to the pistol's lanyard loop.***
Photographs by A. Vanderlinden, courtesy Waclaw Ustupski collection

Polish Production: 1937 Eagle VIS N° 7173

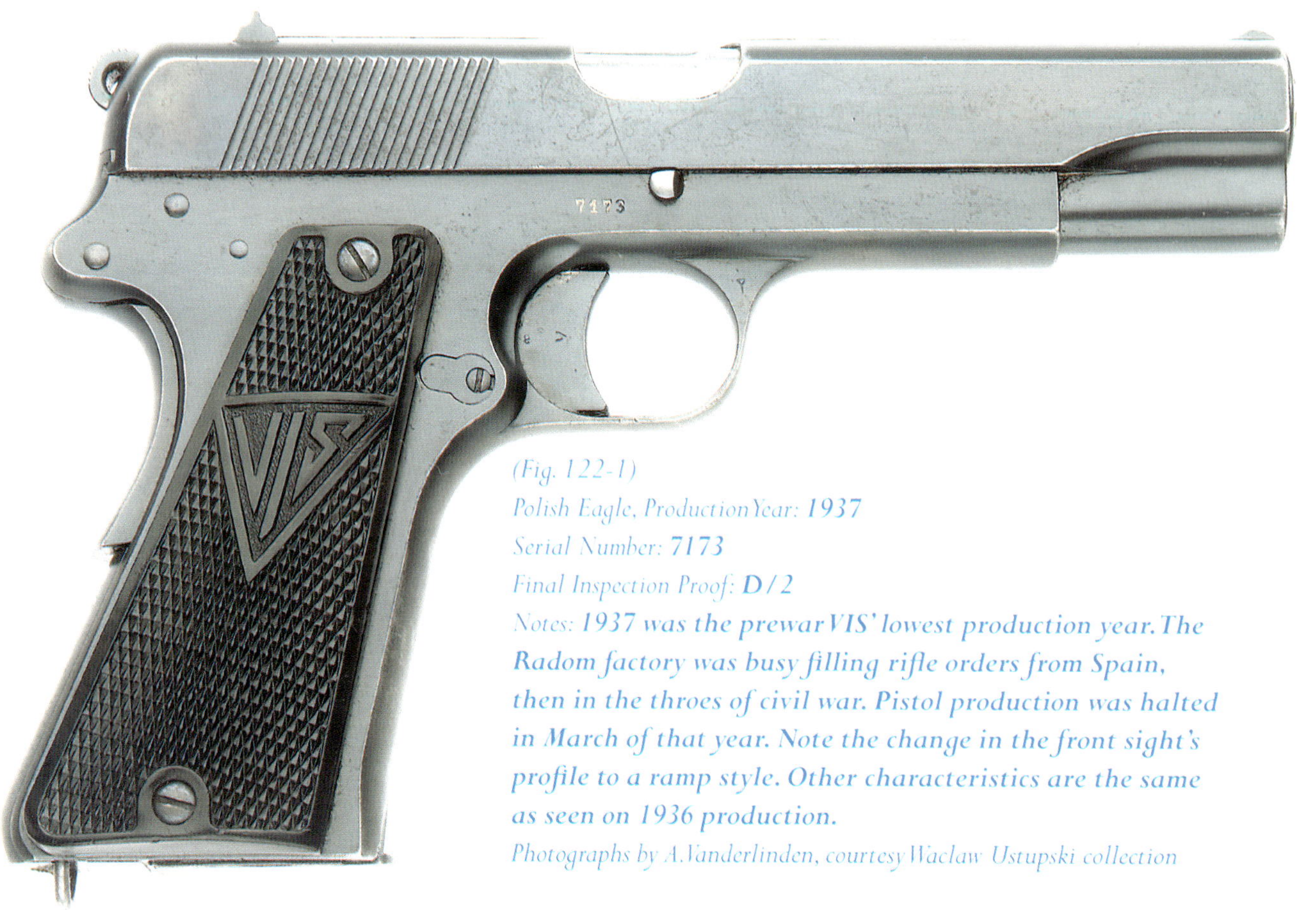

(Fig. 122-1)
Polish Eagle, Production Year: **1937**
Serial Number: **7173**
Final Inspection Proof: **D/2**
Notes: ***1937 was the prewar VIS' lowest production year. The Radom factory was busy filling rifle orders from Spain, then in the throes of civil war. Pistol production was halted in March of that year. Note the change in the front sight's profile to a ramp style. Other characteristics are the same as seen on 1936 production.***
Photographs by A. Vanderlinden, courtesy Waclaw Ustupski collection

Polish Production: 1938 Eagle VIS N° 15219

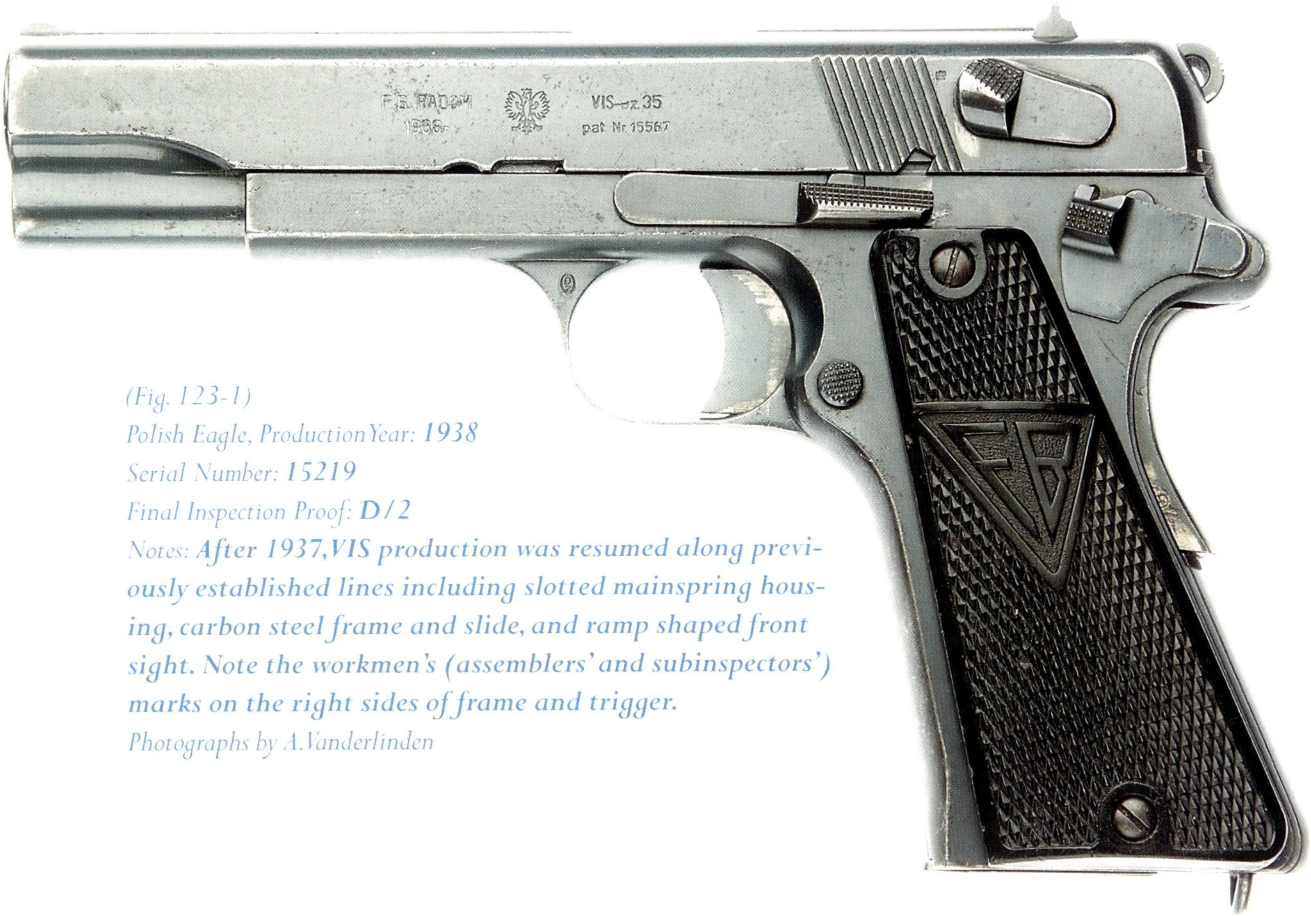

(Fig. 123-1)
Polish Eagle, Production Year: ***1938***
Serial Number: ***15219***
Final Inspection Proof: ***D/2***
Notes: ***After 1937, VIS production was resumed along previously established lines including slotted mainspring housing, carbon steel frame and slide, and ramp shaped front sight. Note the workmen's (assemblers' and subinspectors') marks on the right sides of frame and trigger.***
Photographs by A. Vanderlinden

Polish Production: 1938 Eagle VIS N° 16151

(Fig. 124-1)
Polish Eagle, Production Year: ***1938***
Serial Number: ***16151***
Final Inspection Proof: ***B / 8***
Notes: ***This prewar pistol shows wartime use by the German Navy. The pistol is marked "O4238" on the frontstrap, signifying the OSTSEE (Baltic) fleet. This is a very seldom seen marking. See fig. 149-1 for a NORDSEE marked pistol. This pistol shows purely Polish production markings, with no additional German wartime markings. See the warning on counterfeit Geman Naval markings on page 148.***
Photographs by A. Vanderlinden, courtesy Ron Pinkenburg collection

(Fig. 125-1) Notes: **Polish assemblers / subinspectors' markings on the left side of trigger and slide.**

(Fig. 125-2) Notes: **The Ostsee "O" marking and related number on the pistol's front grip strap.**

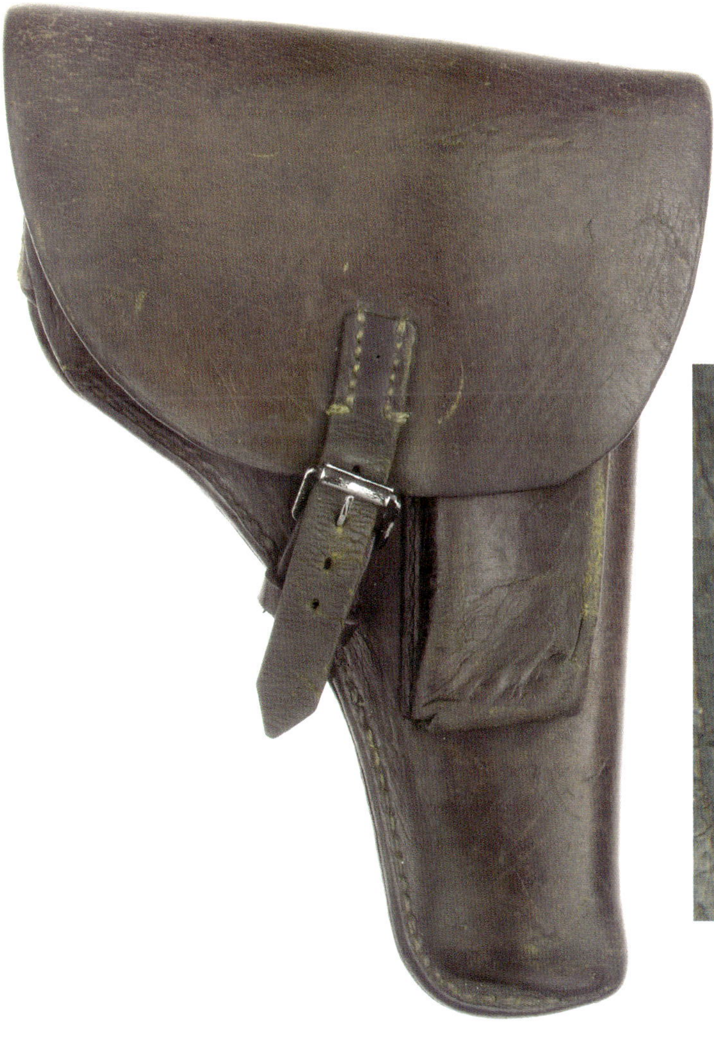

(Fig. 125-3) Notes: **The German Naval "Marine" marking on the pistol's holster. This holster is unusual in having a buckle closure.**

Polish Production: 1938 Eagle VIS N° 16617

(Fig. 126-1)
Polish Eagle, Production Year: ***1937***
Serial Number: ***16617***
Final Inspection Proof: ***B/8***
Notes: ***See page 127***
Photographs by A.Vanderlinden, courtesy Waclaw Ustupski collection

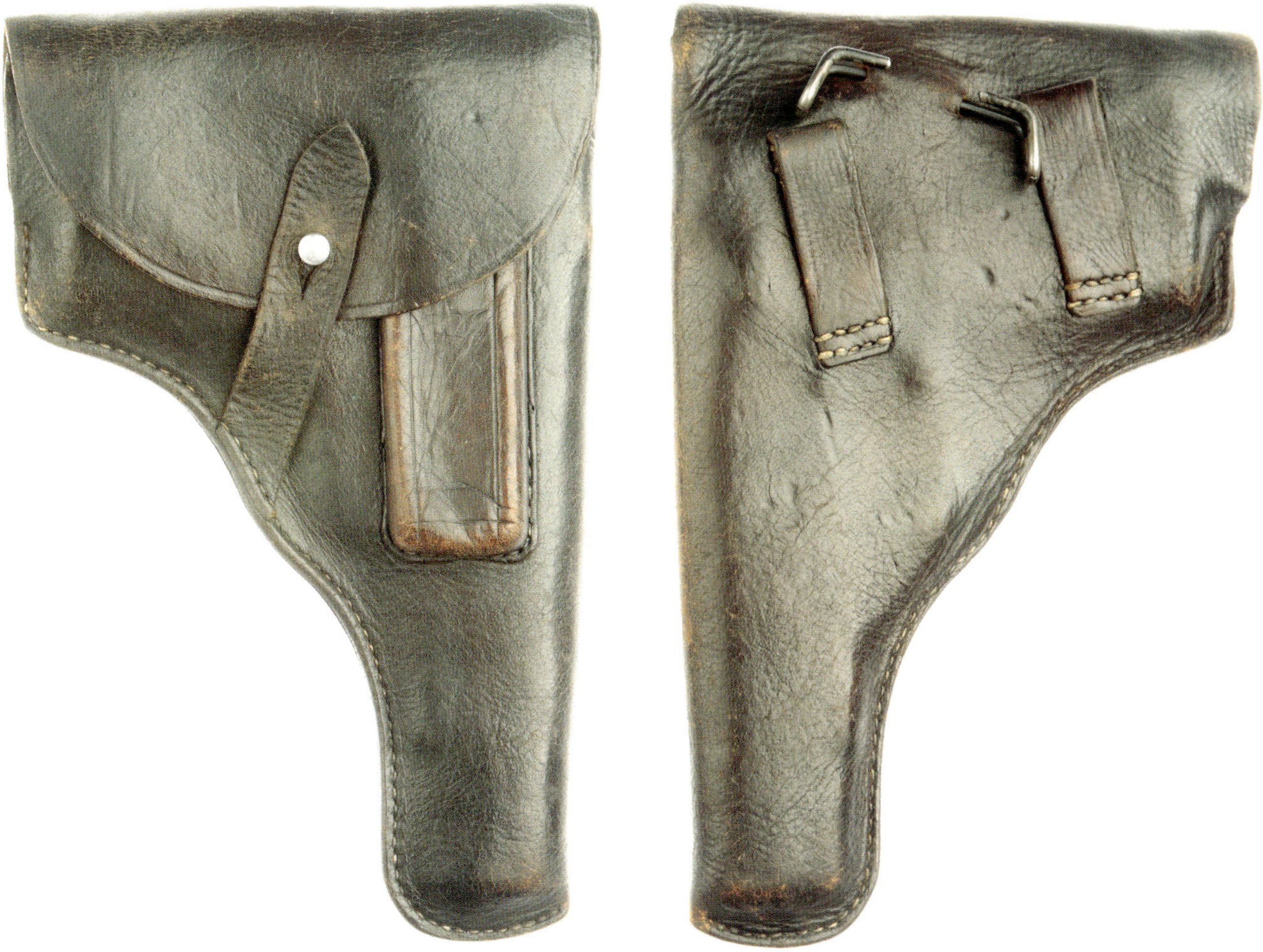

(Fig. 127-1)

Notes: ***This pistol and holster came with provenance documents: The rig was captured by Willard Wycoff, a combat engineer, with the 160th Engineers' Combat Battalion which fought aside the 101st Airborne liberating Western Europe.***

The combat engineers were essential at keeping the American momentum going, members of the 160th were all parachutists and known for getting the job done under fire. Wycoff's unit was known for building the longest floating bridge across the Rhine river while under fire. A feat which earned the 160th a Presidential Citation.

Wycoff and the 160th found themselves in many well known battles including Normandy where they parachuted into St. Mere Eglise, followed by Bastogne in the Battle of the Bulge. Other memorable moments were the liberation of the Orduff concentration camp and their presence in Berchtesgaten, Hitler's retreat. Wycoff earned the Bronze Star for bravery in combat and a Purple Heart for sustained injuries. Somewhere along the way he captured this Radom pistol with non-standard holster. This holster shows a "GI" modification by addition of a U.S. pistol belt hanger.

Polish Production: 1938 Eagle VIS Consecutive Numbers N° 17808 & 17809

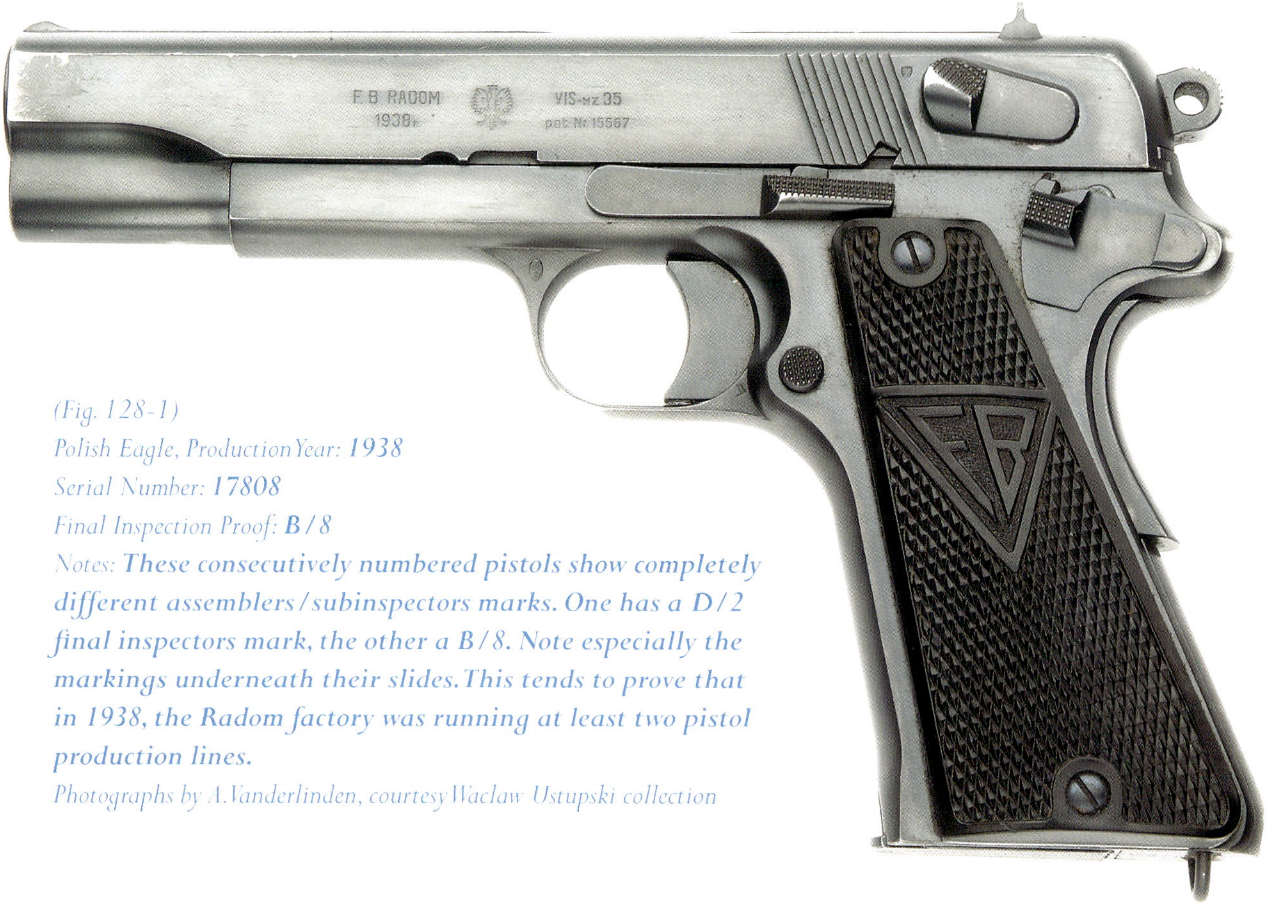

(Fig. 128-1)
Polish Eagle, Production Year: ***1938***
Serial Number: ***17808***
Final Inspection Proof: ***B/8***
Notes: ***These consecutively numbered pistols show completely different assemblers/subinspectors marks. One has a D/2 final inspectors mark, the other a B/8. Note especially the markings underneath their slides. This tends to prove that in 1938, the Radom factory was running at least two pistol production lines.***
Photographs by A. Vanderlinden, courtesy Waclaw Ustupski collection

(Fig. 128-2)
Polish Eagle, Production Year: ***1938***
Serial Number: ***17809***
Final Inspection Proof: ***D/2***
Notes: ***These consecutively numbered pistols show completely different assemblers/subinspectors marks. One has a D/2 final inspectors mark, the other a B/8. Note especially the markings underneath their slides. This tends to prove that in 1938, the Radom factory was running at least two pistol production lines.***
Photographs by A. Vanderlinden, courtesy Waclaw Ustupski collection

VIS
VIS

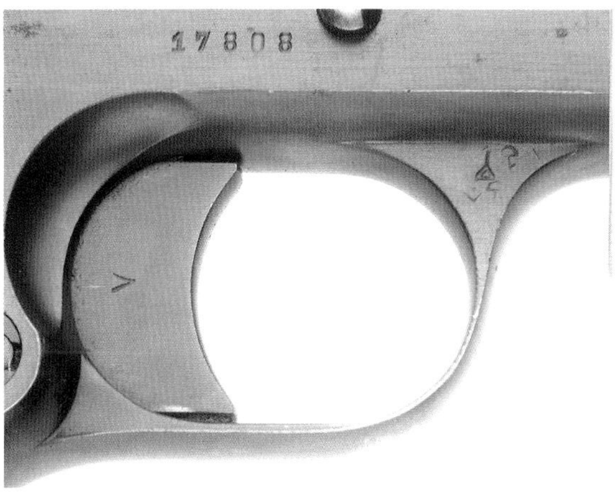
17808

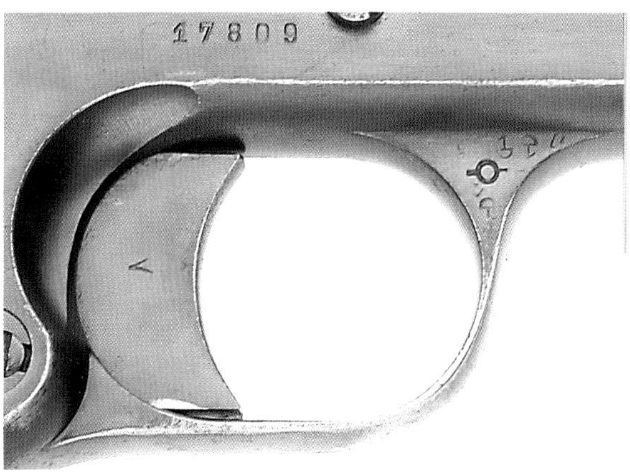
17809

D2

Polish Production: 1938 Eagle VIS Consecutive Numbers N° 17808 & 17809

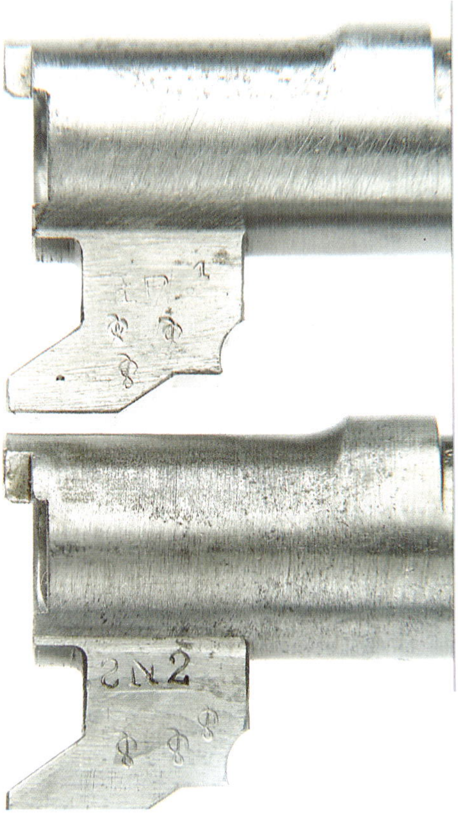

(Fig. 130-1)

Notes: ***Although consecutively numbered, these parts show numerous differences in markings and proofs, indicating that two assembly lines were most likely in place.***

Polish Production: Eagle VIS N° 18374

(Fig. 131-1)
Production Year: ***Unknown***
Serial Number: ***18374***
Final Inspection Proof: ***D/2***
Notes: ***This pistol has a frame number in the 1938 range, and a D/2 final inspectors mark, but a blank slide. It is a possible early German put-together, or a late Polish example of the same. The assemblers/subinspectors' markings are Polish.***
Photographs by A. Vanderlinden, courtesy Ron Pinkenburg collection

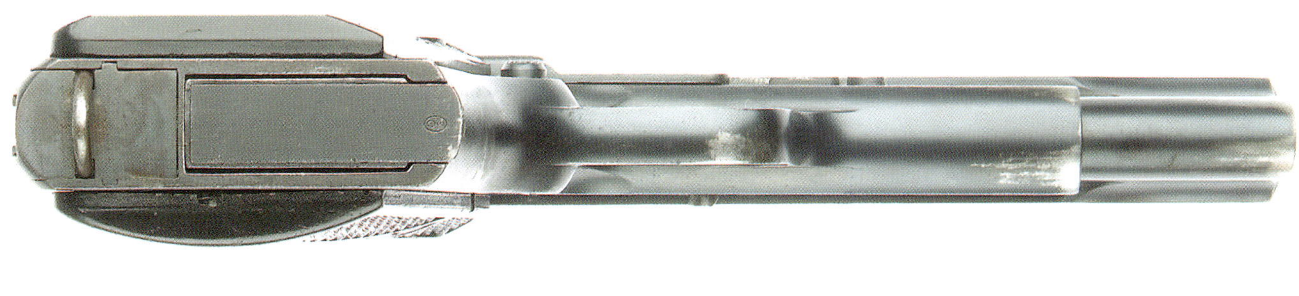

Polish Production: 1939 Eagle VIS N° 29284

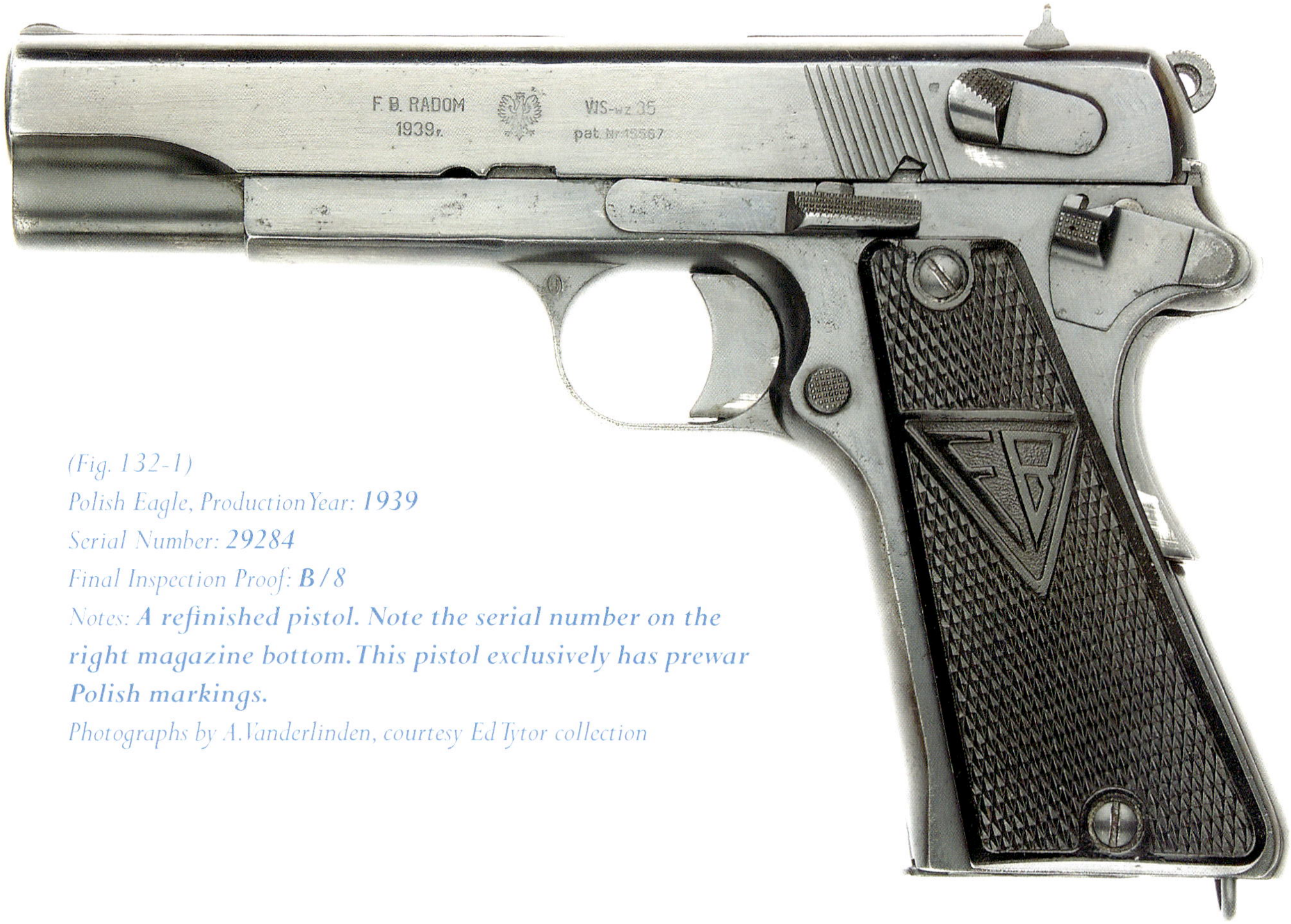

(Fig. 132-1)
Polish Eagle, Production Year: ***1939***
Serial Number: ***29284***
Final Inspection Proof: ***B/8***
Notes: ***A refinished pistol. Note the serial number on the right magazine bottom. This pistol exclusively has prewar Polish markings.***
Photographs by A. Vanderlinden, courtesy Ed Tytor collection

Polish Production: 1939 Eagle VIS Nº 29391

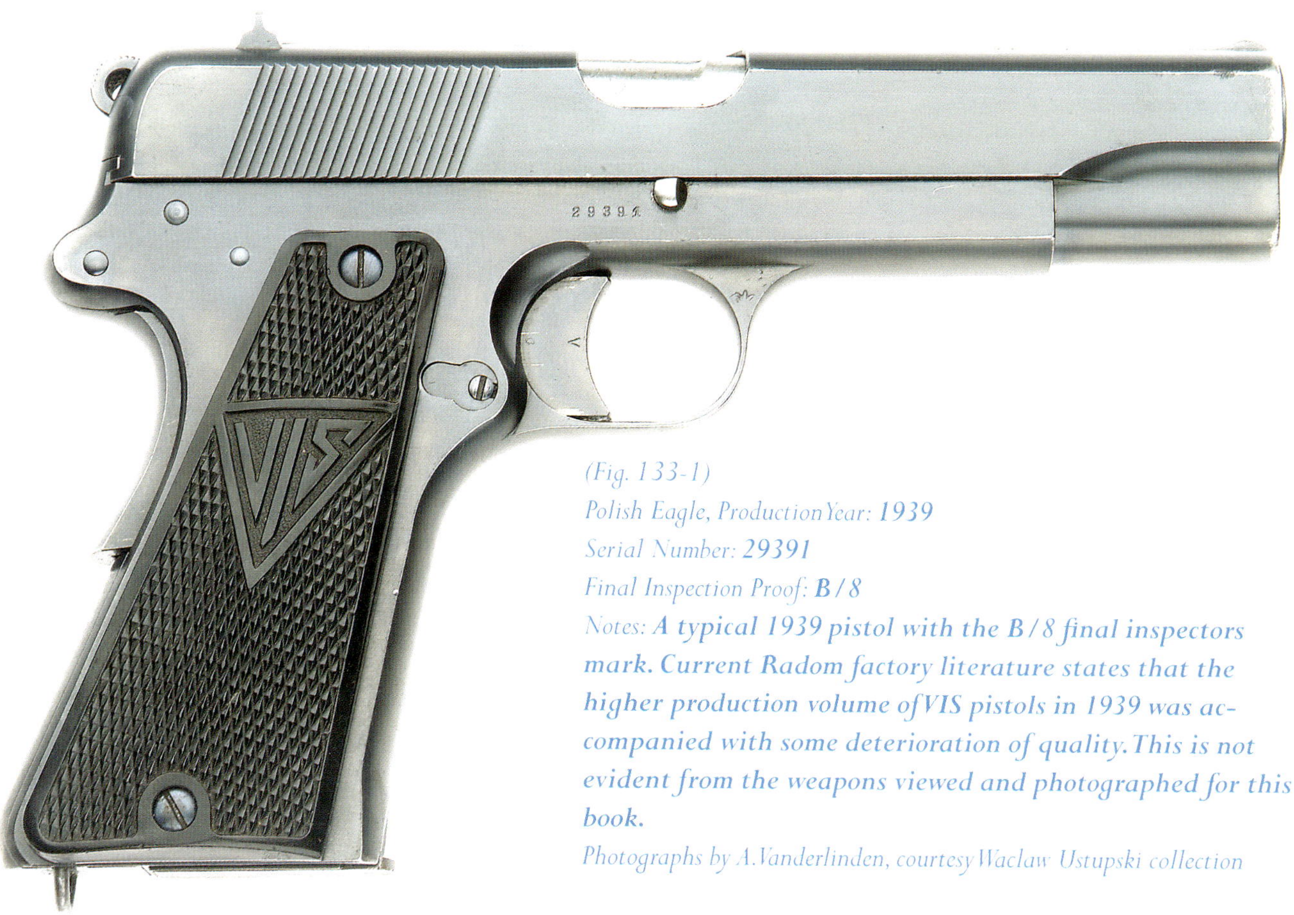

(Fig. 133-1)
Polish Eagle, Production Year: ***1939***
Serial Number: ***29391***
Final Inspection Proof: ***B/8***
Notes: ***A typical 1939 pistol with the B/8 final inspectors mark. Current Radom factory literature states that the higher production volume of VIS pistols in 1939 was accompanied with some deterioration of quality. This is not evident from the weapons viewed and photographed for this book.***
Photographs by A. Vanderlinden, courtesy Waclaw Ustupski collection

Polish Production: 1939 Eagle VIS N° 30234

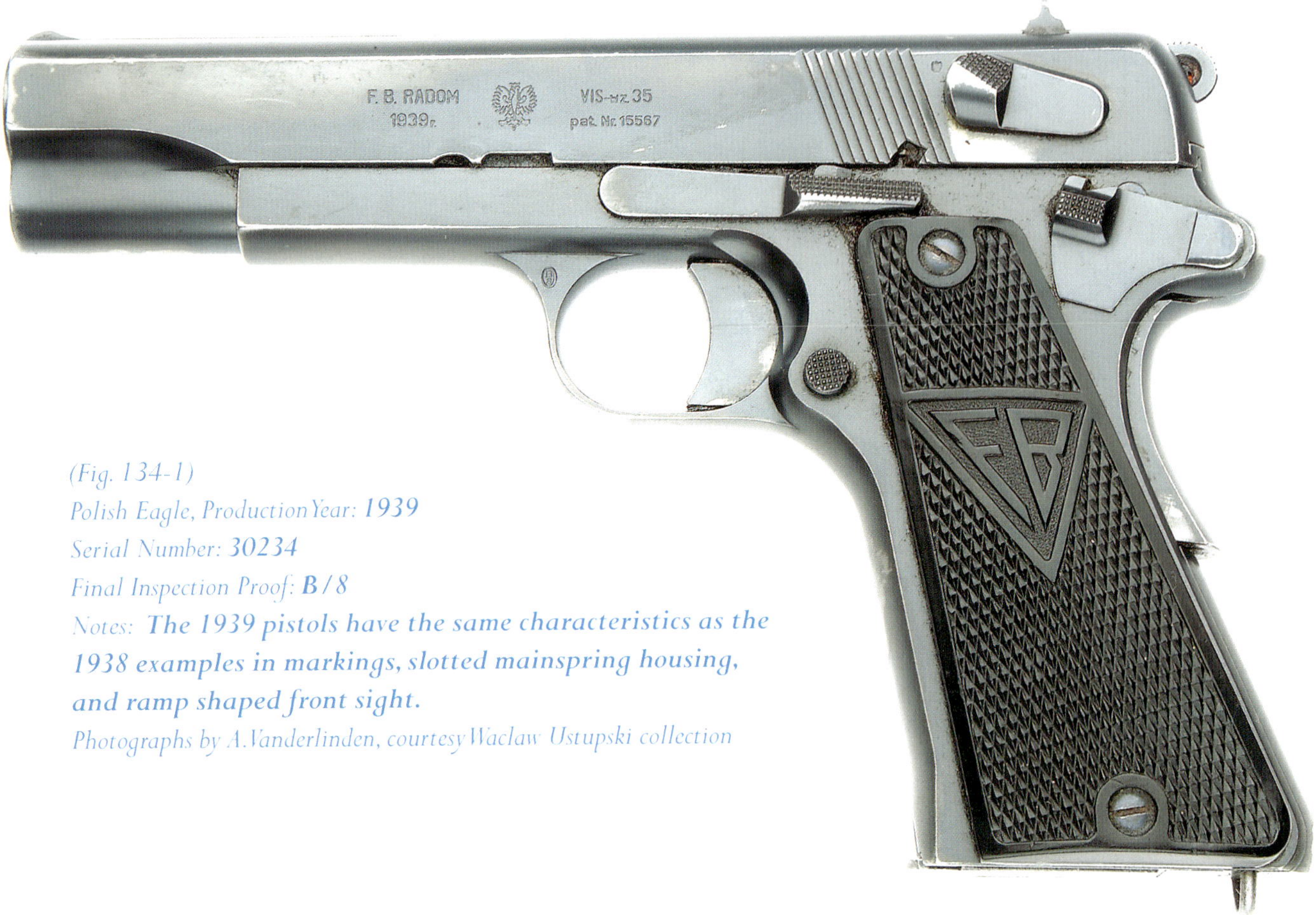

(Fig. 134-1)
Polish Eagle, Production Year: ***1939***
Serial Number: ***30234***
Final Inspection Proof: ***B/8***
Notes: ***The 1939 pistols have the same characteristics as the 1938 examples in markings, slotted mainspring housing, and ramp shaped front sight.***
Photographs by A. Vanderlinden, courtesy Waclaw Ustupski collection

Polish Production: 1939 Eagle VIS N° 34523

(Fig. 135-1)
Polish Eagle, Production Year: ***1939***
Serial Number: ***34523***
Final Inspection Proof: ***D/2***
Notes: ***A finely finished pistol of Polish production with typical characteristics and markings.***
Photographs by A. Vanderlinden, courtesy Waclaw Ustupski collection

Polish Production: 1939 Eagle VIS Nº 35173

(Fig. 136-1)
Polish Eagle, Production Year: ***1939***
Serial Number: ***35173***
Final Inspection Proof: ***D/2***
Notes: ***Another 1939 VIS pistol with typical prewar characteristics.***
Photographs by A.Vanderlinden

Polish Production: 1939 Eagle VIS N° 40416

(Fig. 137-1)
Polish Eagle, Production Year: ***1939***
Serial Number: ***40416***
Final Inspection Proof: ***D/2***
Notes: ***The Delta symbol following the serial number on some prewar VIS pistols is believed to be a factory repair mark. It was most likely applied following repair of the hammer drop mechanism, which is often found to be out of order on these pistols. Note also the presence of a a stylized and mirror-images circled "I" mark above the final inspectors D/2 mark. This may also be a factory repair mark. The "x" marking on the left side of the trigger and the right front triggerguard is also unusual.***
Photographs by A. Vanderlinden, courtesy Waclaw Ustupski collection

Polish Production: Eagle VIS N° 47841

(Fig. 138-1)
Polish Eagle, Production Year: ***1939***
Serial Number: ***Frame 47841, slide 48813***
Final Inspection Proof: ***None***
Notes: ***This is a mismatched pistol with a 1939 era frame and a 1938 dated slide, and no lanyard loop. It is most likely a pistol put together during the German invasion of Poland. The pistol shows no German markings.***
Photographs by A. Vanderlinden, courtesy Waclaw Ustupski collection

Polish Production: Eagle VIS N° 48517

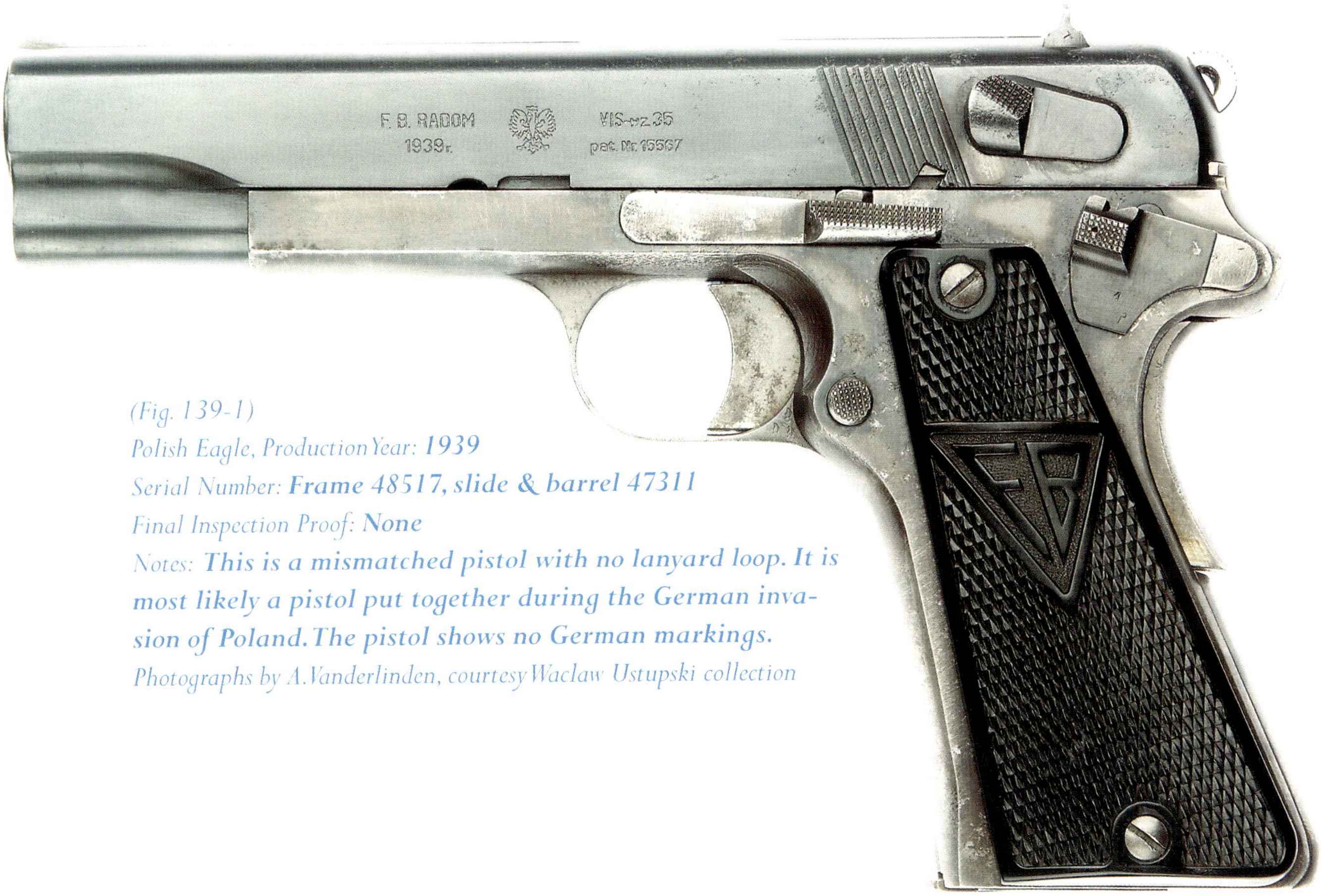

(Fig. 139-1)
Polish Eagle, Production Year: ***1939***
Serial Number: ***Frame 48517, slide & barrel 47311***
Final Inspection Proof: ***None***
Notes: ***This is a mismatched pistol with no lanyard loop. It is most likely a pistol put together during the German invasion of Poland. The pistol shows no German markings.***
Photographs by A.Vanderlinden, courtesy Waclaw Ustupski collection

Polish Production: Eagle VIS N° 48907

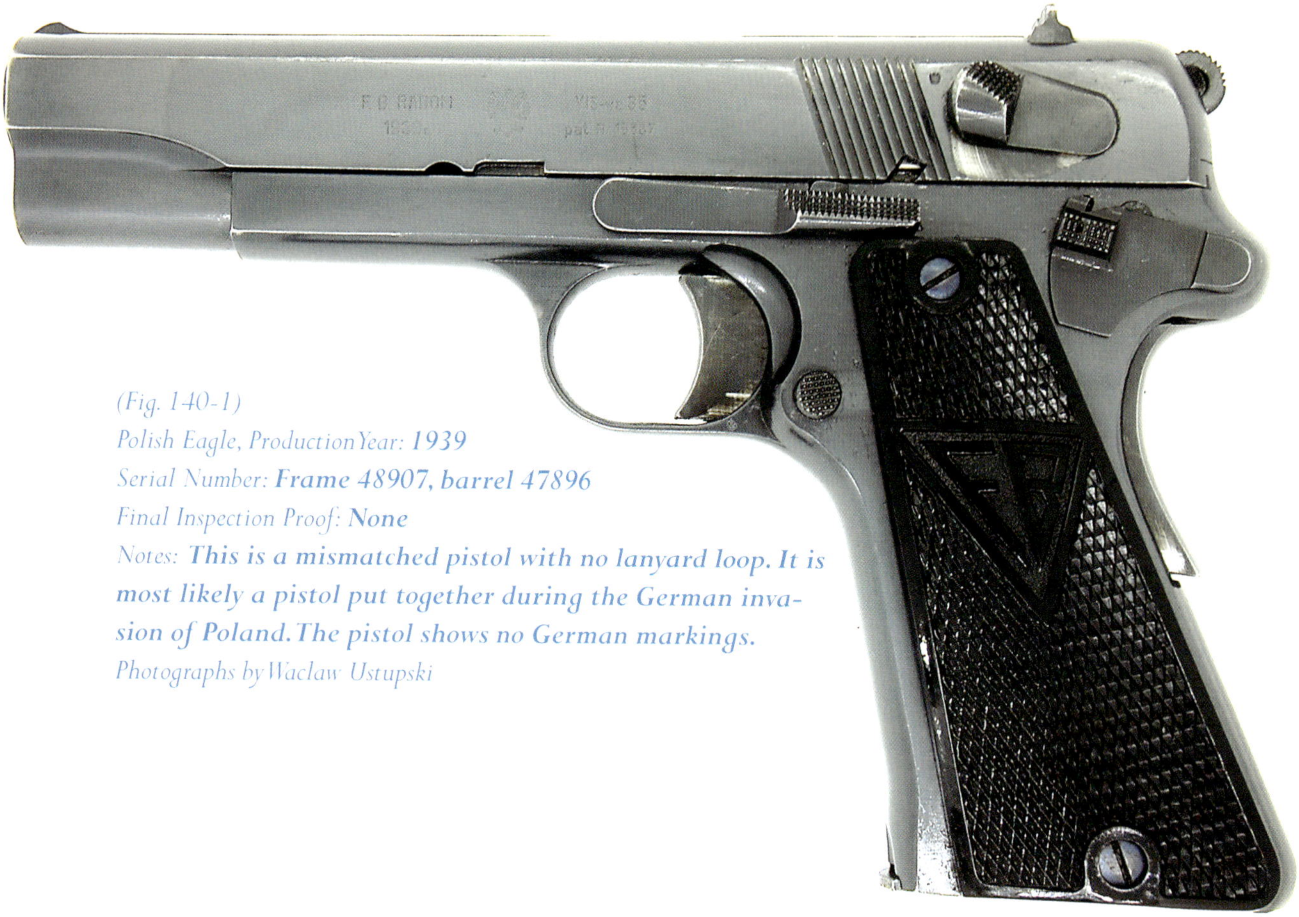

(Fig. 140-1)
Polish Eagle, Production Year: **1939**
Serial Number: **Frame 48907, barrel 47896**
Final Inspection Proof: **None**
Notes: ***This is a mismatched pistol with no lanyard loop. It is most likely a pistol put together during the German invasion of Poland. The pistol shows no German markings.***
Photographs by Waclaw Ustupski

Polish Production: Eagle VIS Nº 49014

(Fig. 141-1)
Polish Eagle, Production Year: ***1939***
Serial Number: ***49014***
Final Inspection Proof: ***None***
Notes: ***A VIS pistol with all typical Polish assemblers/sub-inspectors' markings, with no lanyard loop, and lacking slide markings, a final inspector's marking, or a firing proof. Likely assembled either at the very end of Polish production or during the early occupation period before German production began.***
Photographs by A. Vanderlinden, courtesy Ron Pinkenburg collection

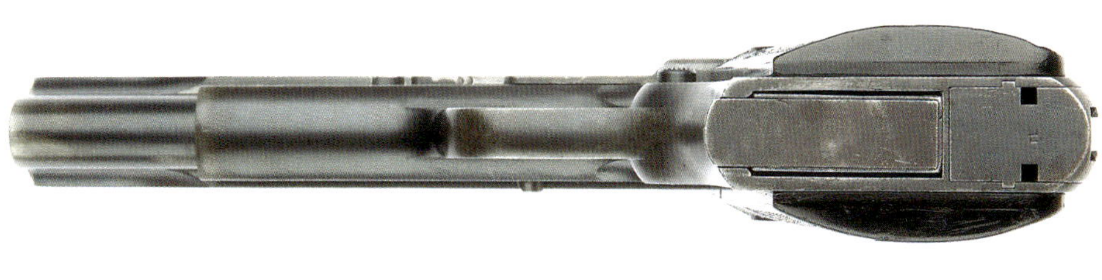

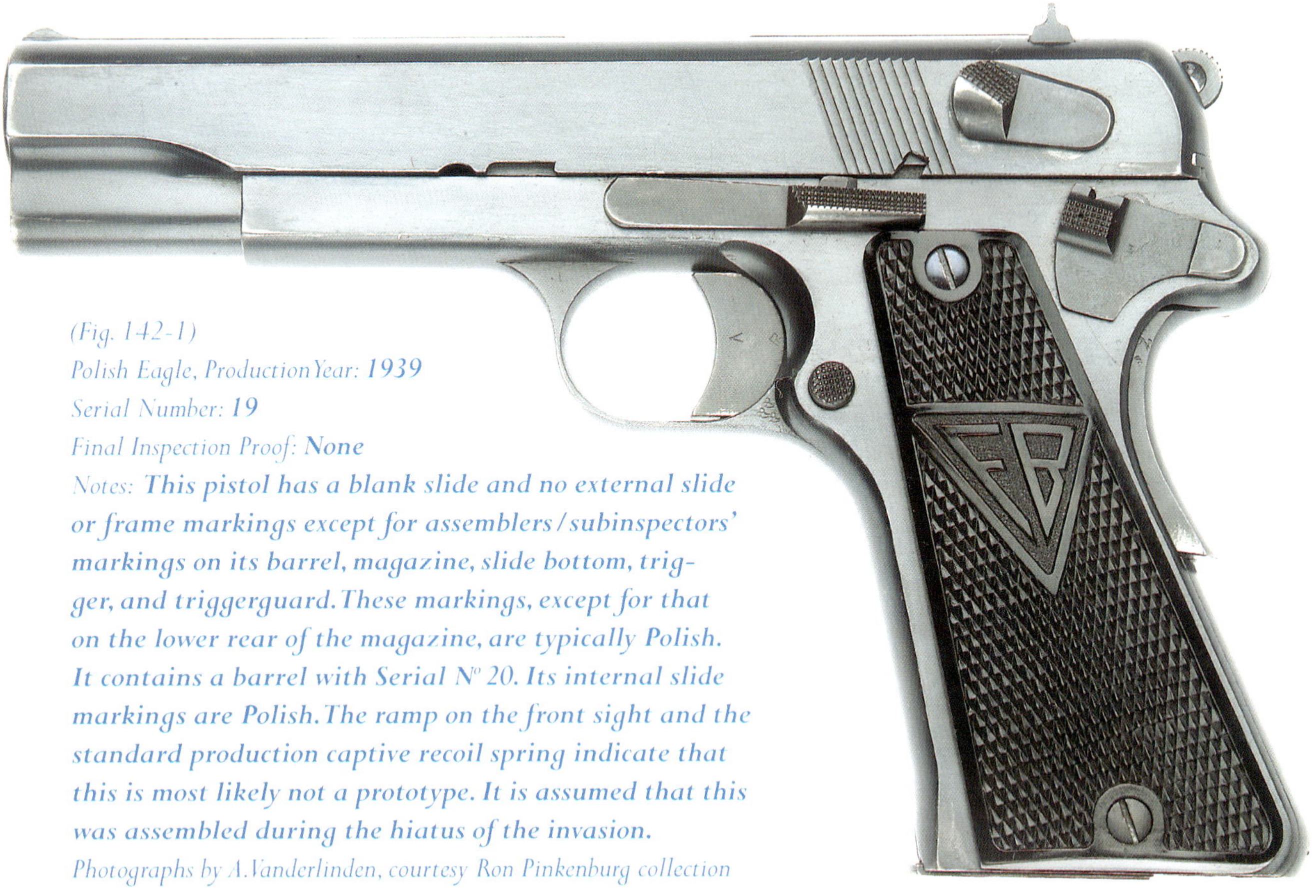

(Fig. 142-1)
Polish Eagle, Production Year: ***1939***
Serial Number: ***19***
Final Inspection Proof: ***None***
Notes: ***This pistol has a blank slide and no external slide or frame markings except for assemblers/subinspectors' markings on its barrel, magazine, slide bottom, trigger, and triggerguard. These markings, except for that on the lower rear of the magazine, are typically Polish. It contains a barrel with Serial N° 20. Its internal slide markings are Polish. The ramp on the front sight and the standard production captive recoil spring indicate that this is most likely not a prototype. It is assumed that this was assembled during the hiatus of the invasion.***
Photographs by A. Vanderlinden, courtesy Ron Pinkenburg collection

(Fig. 143-1) Notes: **The slide bottom, right barrel lug, and bottom rear of the magazine from pistol serial number 19**

Photography Gallery II

Occupation Production at Radom & Steyr
Pre-Alphabet and First Alphabet Prefix Series

Pistole 645 (p) – poln 35 – Pistolet „Vis" wz-35

Kennblätter fremden Geräts

D 50/1
Polen

dtsch Kennummer
645 01 (p)

poln Ben: Pistolet „Vis" wz—35
Abk:

dtsch Ben: **Pistole 645 (p)**
wo nötig erläutern mit } - - poln „Vis" 35 - -

Abk: P 645 (p)

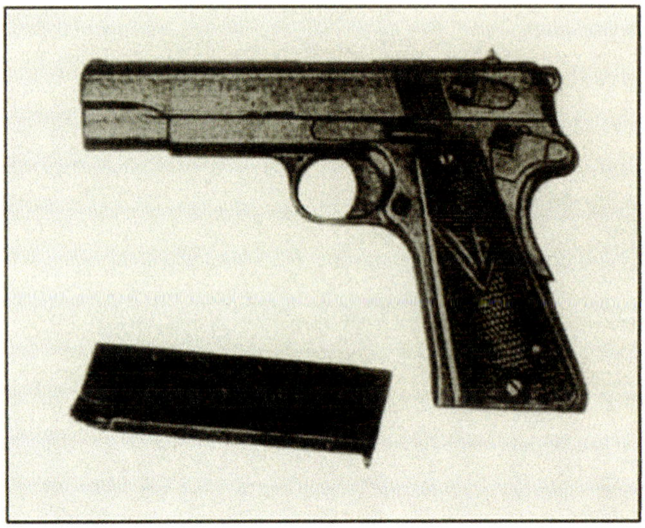

Kennzeichen
An der linken Seite des Gleitstückes poln Firmen- und Hoheitszeichen und Bezeichnung „Vis" wz 35"

Bauart
Rückstoßlader mit Kipplaufverriegelung, ähnlich System „Colt"

Kaliber	9 mm
Gewicht	1,02 kg
Länge	210 mm
Lauflänge	120 mm
Zahl der Züge	6
Drallrichtung	rechts

Visier
Standvisier

Patronenzuführung
Griffstückmagazin für 8 Patronen

Patronen
9 mm Pist Patr „Parabellum". Dtsch. Pist Patr 08 können verschossen werden

Feuergeschwindigkeit
12 bis 16 Schuß/Min

Gebrauchsentfernung
bis 50 m

Occupation Production: VIS N° 0108

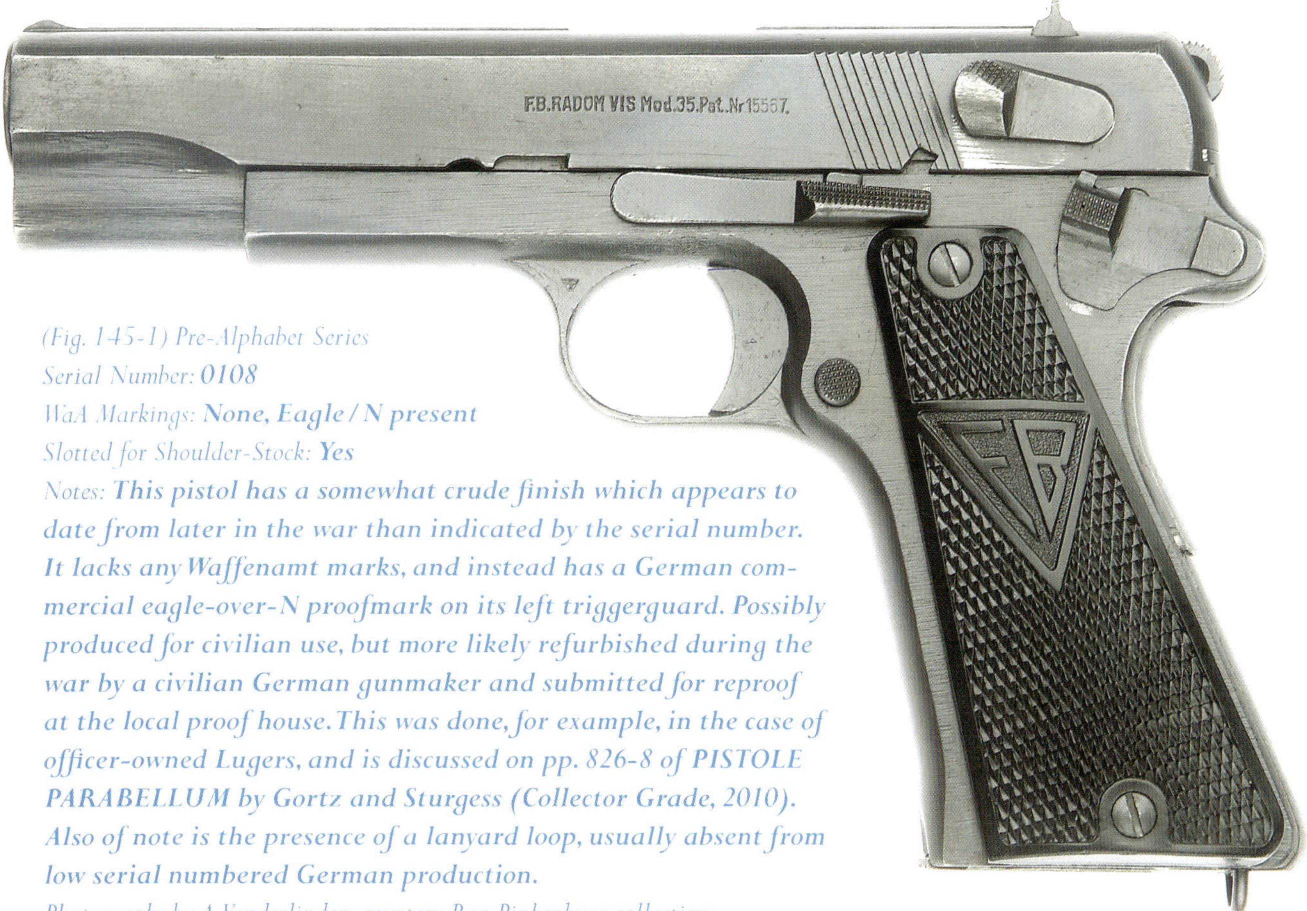

(Fig. 145-1) Pre-Alphabet Series
Serial Number: ***0108***
WaA Markings: ***None, Eagle / N present***
Slotted for Shoulder-Stock: ***Yes***
Notes: ***This pistol has a somewhat crude finish which appears to date from later in the war than indicated by the serial number. It lacks any Waffenamt marks, and instead has a German commercial eagle-over-N proofmark on its left triggerguard. Possibly produced for civilian use, but more likely refurbished during the war by a civilian German gunmaker and submitted for reproof at the local proof house. This was done, for example, in the case of officer-owned Lugers, and is discussed on pp. 826-8 of PISTOLE PARABELLUM by Gortz and Sturgess (Collector Grade, 2010). Also of note is the presence of a lanyard loop, usually absent from low serial numbered German production.***
Photographs by A.Vanderlinden, courtesy Ron Pinkenburg collection

(Fig. 146-1) Pre-Alphabet Series
Serial Number: ***204***
WaA Markings: ***WaA77 on slide***
Slotted for Shoulder-Stock: ***Yes***
Notes: ***This is an early German pistol. It lacks the Radom WAa77 Waffenamt marking on its frame. It is notable for the striking plum color of its finish. This color variation is known to occur in this serial number range, and was caused by the presence of variant chemicals, or impurities, in the bluing salts, or by a variation in the bluing salts' temperature. The lack of a lanyard loop is common in very early German occupation production.***
Photographs by A. Vanderlinden, courtesy Ron Pinkenburg collection

Occupation Production: VIS N° 1278

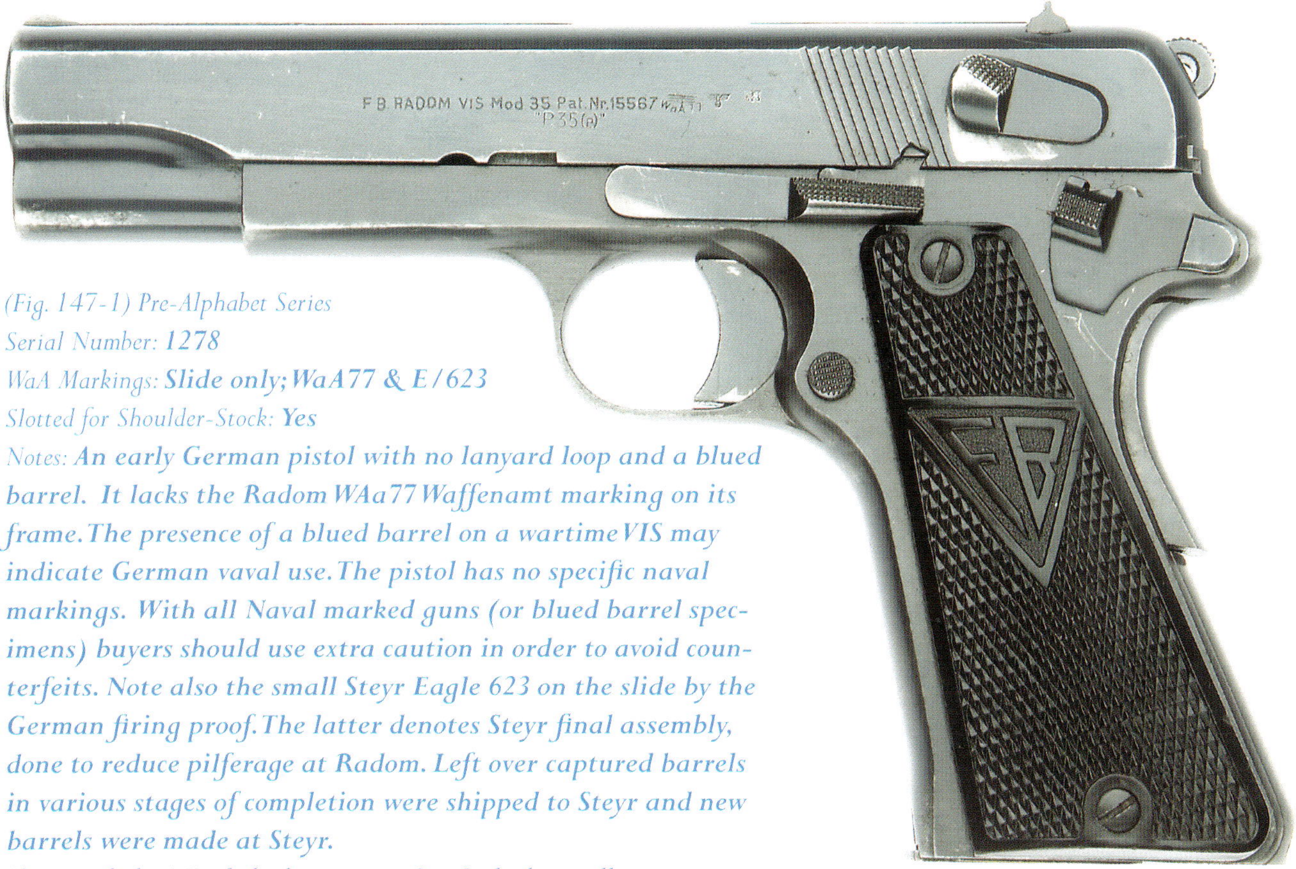

(Fig. 147-1) Pre-Alphabet Series
Serial Number: ***1278***
WaA Markings: ***Slide only; WaA77 & E/623***
Slotted for Shoulder-Stock: ***Yes***
Notes: ***An early German pistol with no lanyard loop and a blued barrel. It lacks the Radom WAa77 Waffenamt marking on its frame. The presence of a blued barrel on a wartime VIS may indicate German vaval use. The pistol has no specific naval markings. With all Naval marked guns (or blued barrel specimens) buyers should use extra caution in order to avoid counterfeits. Note also the small Steyr Eagle 623 on the slide by the German firing proof. The latter denotes Steyr final assembly, done to reduce pilferage at Radom. Left over captured barrels in various stages of completion were shipped to Steyr and new barrels were made at Steyr.***
Photographs by A. Vanderlinden, courtesy Ron Pinkenburg collection

Occupation Production: VIS Nº 8441

(Fig. 148-1) Pre-Alphabet Series
Serial Number: ***8441***
WaA Markings: ***WaA77 & E/623***
Slotted for Shoulder-Stock: ***Yes***
Notes: ***This is an early German pistol with the early Radom WaA77 marking on its slide and frame. It has a blued barrel, and an eagle/swastika over M (MARINE) on the left frame; a German Navy marking. It also has the letter "N" on the front of its frame, signifying the "NORDSEE" (North Sea) fleet. Pistols bearing an "O" marking instead of "N" were issued to the "OSTSEE" (Baltic Sea) fleet. All German Naval-marked pistols are rare, ones with the "O" marking especially so. Buyers should use caution as counterfeits do exist.***
Photographs by A.Vanderlinden, courtesy Ron Pinkenburg collection

(Fig. 149-1) Notes: ***3/4 view from the rear of serial Nº 8441, and 3/4 view of the markings on the front gripstrap of serial Nº 8441***
Photographs by A.Vanderlinden, courtesy Ron Pinkenburg collection

Occupation Production: VIS Nº 8539

(Fig. 150-1) Pre-Alphabet Series
Serial Number: ***8539***
WaA Markings: ***WaA77 , E/623***
Slotted for Shoulder-Stock: ***Yes***

Notes: ***An early German production pistol with typical characteristics. It lacks a lanyard loop, and features the early war P35(P) and WaA77 marks. Note also the small Steyr E/623 on the slide by the German firing proof. The latter denotes Steyr final assembly, done to reduc pilferage at Radom. Left over captured barrels in various stages of completion were shipped t Steyr and new barrels were made at Steyr.***

Photographs by A.Vanderlinden, courtesy W. Ustupski

Occupation Production: VIS N° 9591

(Fig. 151-1) Pre-Alphabet Series
Serial Number: ***9591***
WaA Markings: ***WaA77 , E/623***
Slotted for Shoulder-Stock: ***Yes***
Notes: ***This is an early German pistol with the early Radom WaA77 marking on its slide and frame. It has a blued barrel, and an eagle/swastika over M (MARINE) on the left frame- a German Navy marking. It also has the letter N on the front of its frame, signifying the "NORDSEE" (North Sea) fleet. The British proofs indicate a capture piece that was brought back to England after the war.***
Photographs courtesy of Waclaw Ustupski, from the Steve Vandtke collection

Occupation Production: VIS N° 10150

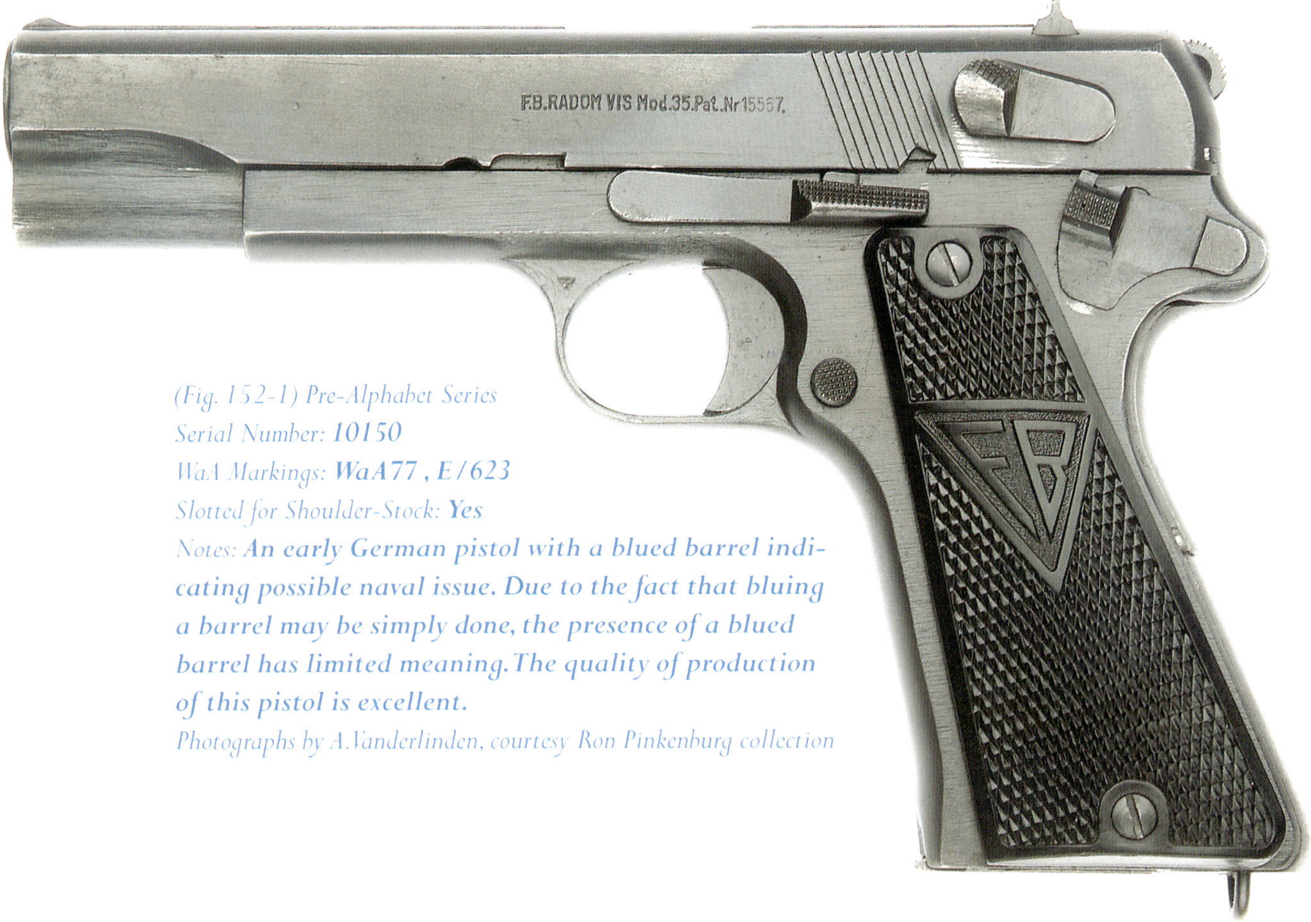

(Fig. 152-1) Pre-Alphabet Series
Serial Number: ***10150***
WaA Markings: ***WaA77 , E/623***
Slotted for Shoulder-Stock: ***Yes***
Notes: ***An early German pistol with a blued barrel indicating possible naval issue. Due to the fact that bluing a barrel may be simply done, the presence of a blued barrel has limited meaning. The quality of production of this pistol is excellent.***
Photographs by A.Vanderlinden, courtesy Ron Pinkenburg collection

Occupation Production: VIS N° 10759

(Fig. 153-1) Pre-Alphabet Series

Serial Number: ***10759***

WaA Markings: ***WaA77 , E/623***

Slotted for Shoulder-Stock: ***Yes***

Notes: ***This pistol has a 1939 dated Polish Eagle slide, but has a German Eagle firing proof, and Radom and Steyr Waffenamt/Eagle marks. It also has a blued barrel indicating possible naval issue. The prewar slide is an indicator of the various stages of manufacture the parts were in when assembled by the Germans. Note the longer than usual lanyard loop, quite likely added after leaving the factory.***

Photographs by A.Vanderlinden, courtesy Ron Pinkenburg collection

Occupation Production: VIS N° 11150

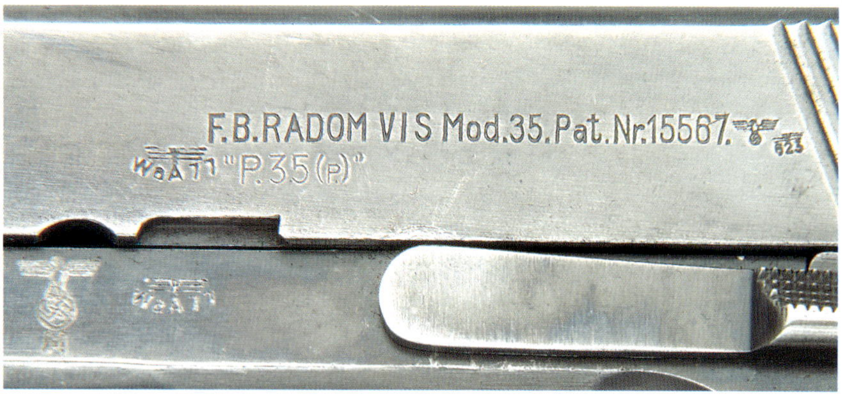

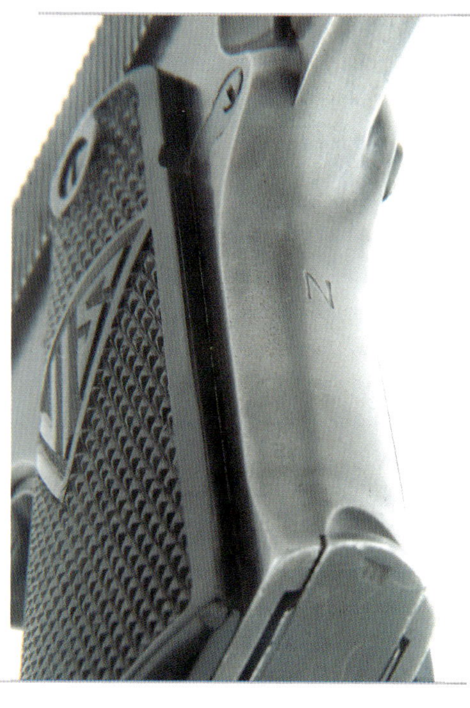

(Fig. 154-1) Pre-Alphabet Series
Serial Number: ***11150***
WaA Markings: ***WaA77 , E/623***
Slotted for Shoulder-Stock: ***Yes***
Notes: ***An early German pistol with naval markings. It also has a blued barrel. A rare pistol, but collectors must beware of counterfeits.***
Photographs by A.Vanderlinden, courtesy Waclaw Ustupski collection

Occupation Production: First alphabet prefix VIS N° A1987

(Fig. 155-1) First Alphabet Series
Serial Number: **A1987**
WaA Markings: **WaA77 , E/623**
Slotted for Shoulder-Stock: **Yes**
Notes: ***This pistol has no lanyard loop and a bright barrel; not blued. It does have German naval (M- Marine) markings, and the letter "N" on the front of its frame, signifying the "NORDSEE" (North Sea) fleet. The lack of a blued barrel seems to call the "blued barrel, equals, Navy" theory into question, but at the same time bolsters the authenticity of this pistol.***
Photographs by A.Vanderlinden, courtesy Ed Tytor collection

Occupation Production: First alphabet prefix VIS N° A7126

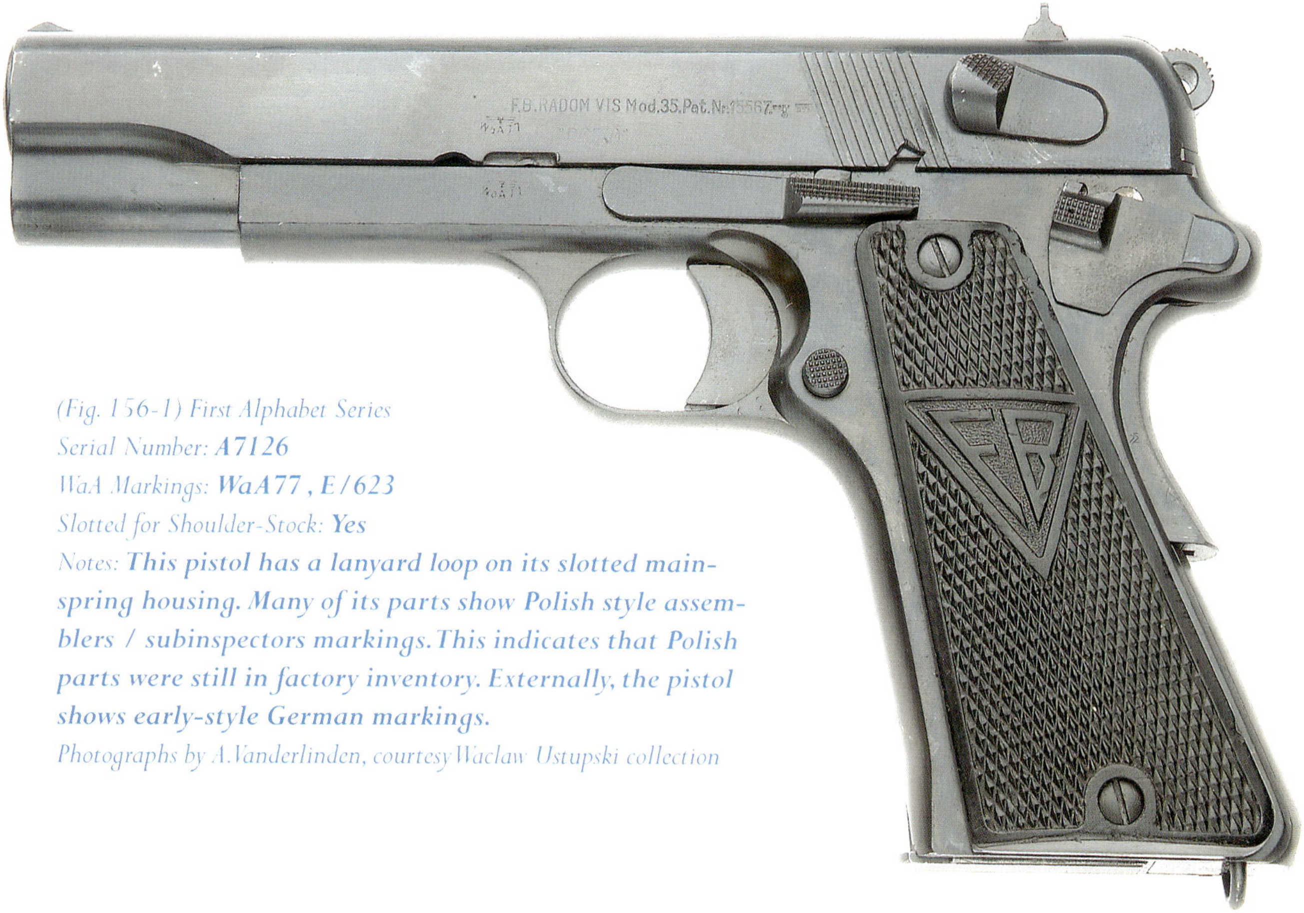

(Fig. 156-1) First Alphabet Series
Serial Number: ***A7126***
WaA Markings: ***WaA77 , E/623***
Slotted for Shoulder-Stock: ***Yes***
Notes: ***This pistol has a lanyard loop on its slotted main-spring housing. Many of its parts show Polish style assemblers / subinspectors markings. This indicates that Polish parts were still in factory inventory. Externally, the pistol shows early-style German markings.***
Photographs by A.Vanderlinden, courtesy Waclaw Ustupski collection

Occupation Production: First alphabet prefix VIS N° B3192

(Fig. 157-1) First Alphabet Series
Serial Number: ***B3192***
WaA Markings: ***WaA77 , E/623***
Slotted for Shoulder-Stock: ***Yes***
Notes: ***This example has one of the last slotted mainspring housings and is equipped with a lanyard loop. Note the Ersatz blackened wooden replacement grip-panel.***
Photographs by Anthony Vanderlinden

Occupation Production: First alphabet prefix VIS N° B3318

(Fig. 158-1) First Alphabet Series
Serial Number: ***B3318***
WaA Markings: ***WaA77 , E/623***
Slotted for Shoulder-Stock: ***No***
Notes: ***This pistol still shows early-style German markings, but lacks the stock slot in its mainspring housing. More war-time parts are being used.***
Photographs by A.Vanderlinden, courtesy Waclaw Ustupski collection

Occupation Production: First alphabet prefix VIS N° C6739

(Fig. 159-1) First Alphabet Series

Serial Number: ***C6739***

WaA Markings: ***WaA77 , E/623***

Slotted for Shoulder-Stock: ***No***

Notes: ***A pistol with early-style P35(p) German markings and no stock slot. The Number 6178 on its frame is an unknown inventory / arsenal number. This pistol has black and brown Bakelite grips.***

Photographs by A. Vanderlinden, courtesy Waclaw Ustupski collection

Occupation Production: First alphabet prefix VIS N° D4475

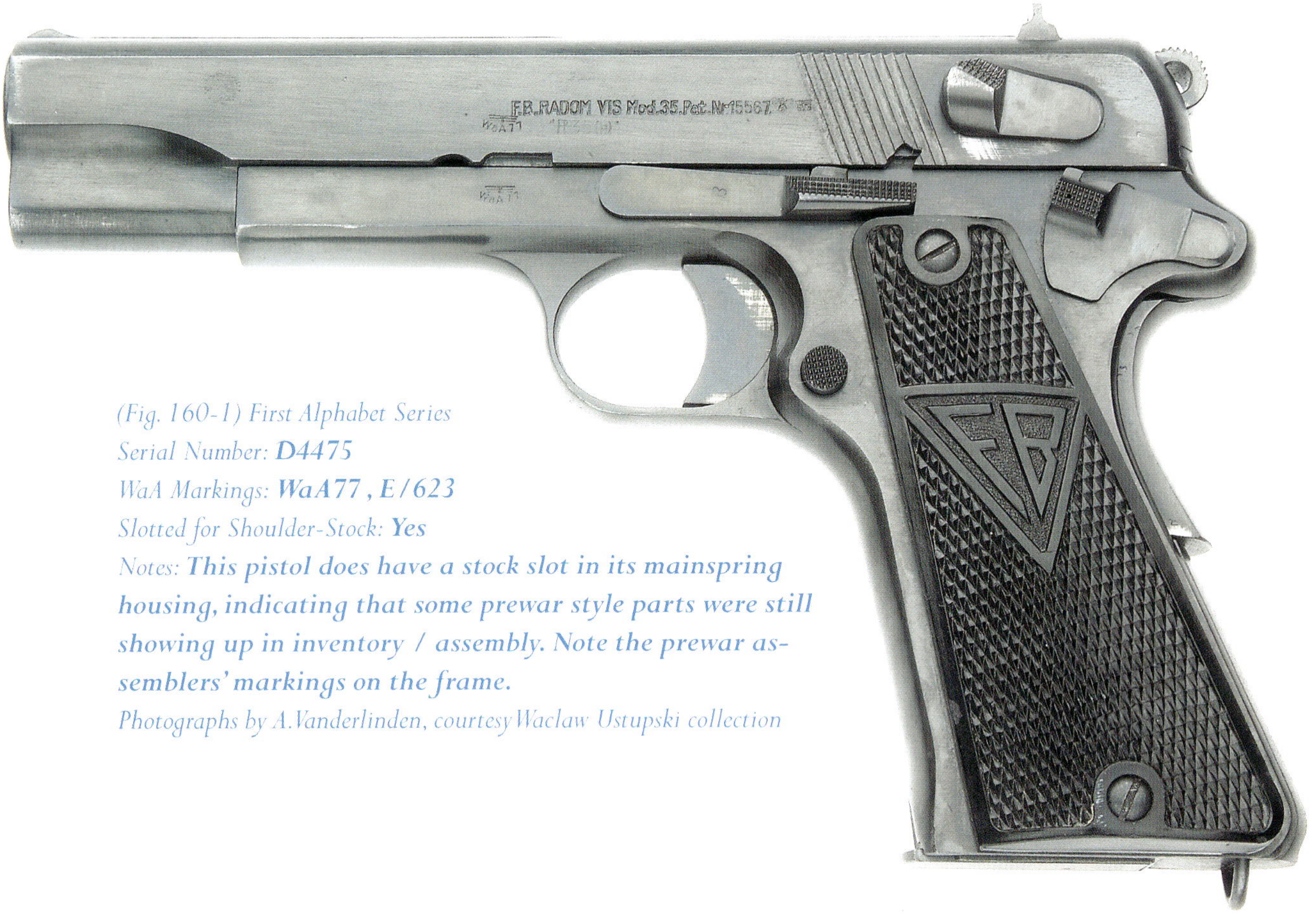

(Fig. 160-1) First Alphabet Series
Serial Number: **D4475**
WaA Markings: **WaA77 , E/623**
Slotted for Shoulder-Stock: **Yes**
Notes: ***This pistol does have a stock slot in its mainspring housing, indicating that some prewar style parts were still showing up in inventory / assembly. Note the prewar assemblers' markings on the frame.***
Photographs by A.Vanderlinden, courtesy Waclaw Ustupski collection

Occupation Production: First alphabet prefix VIS N° D6566

(Fig. 161-1) First Alphabet Series
Serial Number: ***D6566***
WaA Markings: ***WaA77 , E/623***
Slotted for Shoulder-Stock: ***No***
Notes: ***A U.S. GI captured war souvenir. The quality of fit and finish is still very good. The hanging tag attached to the lanyard loop includes the U.S. soldier's name, rank, and Army Service Number***
Photographs by A.Vanderlinden, courtesy Ron Pinkenburg collection

Occupation Production: First alphabet prefix VIS N° D6877

(Fig. 162-1) First Alphabet Series
Serial Number: ***D6877***
WaA Markings: ***WaA77 , E/623***
Slotted for Shoulder-Stock: ***No***
Notes: ***Note that the fit and finish on these pistols is still very good.***
Photographs by A.Vanderlinden, courtesy Waclaw Ustupski collection

Occupation Production: First alphabet prefix VIS Nº D8102

(Fig. 163-1) First Alphabet Series
Serial Number: ***D8102***
WaA Markings: ***WaA77 , E/623***
Slotted for Shoulder-Stock: ***No***
Notes: ***This pistol is also a probable GI war souvenir. It sports Plexiglas grips, which most often were used to protect photographs of wives, family, girlfriends, or pinup girls.***
Photographs by A.Vanderlinden, courtesy Waclaw Ustupski collection

Occupation Production: First alphabet prefix VIS N° D9073

(Fig. 164-1) First Alphabet Series

Serial Number: ***D9073***

WaA Markings: ***WaA77 , E/623***

Slotted for Shoulder-Stock: ***No***

Notes: ***This pistol has a checkered wood right grip, most probably a wartime replacement, Note also the beta character on its slide latch; a subinspectors mark often seen on pistols from this phase of wartime production.***

Photographs by A.Vanderlinden, courtesy Waclaw Ustupski collection

Occupation Production: First alphabet prefix VIS N° E1129

(Fig. 165-1) First Alphabet Series
Serial Number: ***E1129***
WaA Markings: ***WaA77 , E/623***
Slotted for Shoulder-Stock: ***Yes***
Notes: ***This pistol shows some deterioration in quality of finish It does have a slotted mainspring housing.The right grip is wood, with an unchecked triangle but no markings. It appears to be a homemade replacement.***
Photographs by A.Vanderlinden, courtesy Waclaw Ustupski collection

Occupation Production: First alphabet prefix VIS N° E6783

(Fig. 166-1) First Alphabet Series

Serial Number: ***E6783***

WaA Markings: ***WaA77 on frame only***

Slotted for Shoulder-Stock: ***No***

Notes: ***This pistol has a German produced and marked frame mated to a 1938 Polish Eagle slide. There are no German era markings on the slide. This may be a pistol assembled from spare parts.***

Photographs by A. Vanderlinden, courtesy Waclaw Ustupski collection

Occupation Production: First alphabet prefix VIS N° F9077

(Fig. 167-1) First Alphabet Series
Serial Number: **F9077**
WaA Markings: **WaA77 , E/623**
Slotted for Shoulder-Stock: **No**
Notes: **This pistol still shows early German style markings over a rougher finish.**
Photographs by A.Vanderlinden, courtesy Waclaw Ustupski collection

Occupation Production: First alphabet prefix VIS N° G1739

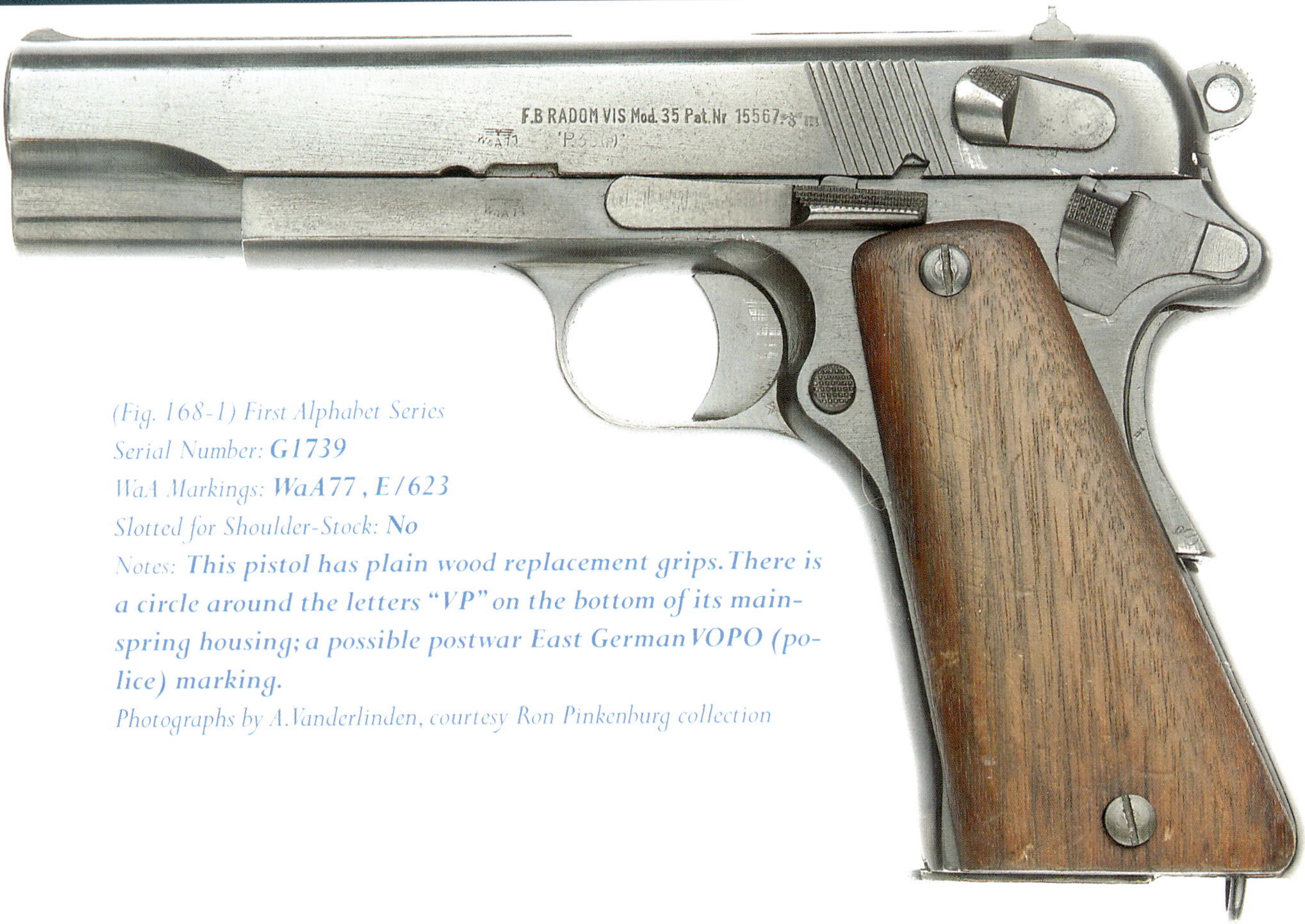

(Fig. 168-1) First Alphabet Series
Serial Number: **G1739**
WaA Markings: **WaA77 , E/623**
Slotted for Shoulder-Stock: **No**
Notes: **This pistol has plain wood replacement grips. There is a circle around the letters "VP" on the bottom of its main-spring housing; a possible postwar East German VOPO (police) marking.**
Photographs by A.Vanderlinden, courtesy Ron Pinkenburg collection

Occupation Production: First alphabet prefix VIS Nº G6567

(Fig. 169-1) First Alphabet Series
Serial Number: **G6567**
WaA Markings: **WaA77 , E/623**
Slotted for Shoulder-Stock: **No**
Notes: ***This is a relatively early German pistol with British proofmarks. Such marks indicate postwar sale in or import to or from the United Kingdom.The number "65" on the right side of the trigger is likely postwar, and is unexplained, although it repeats the last two digits of the serial number.***
Photographs by A.Vanderlinden, courtesy Ron Pinkenburg collection

Occupation Production: First alphabet prefix VIS N° G7160

(Fig. 170-1) First Alphabet Series
Serial Number: ***G7160***
WaA Markings: ***WaA77 , E/623, E/N***
Slotted for Shoulder-Stock: ***No***
Notes: ***This pistol has all the early German military markings, but also has the German wartime commercial "Eagle / N" proofmark on the left trigger guard. A possible wartime repair in a commercial shop (see also fig. 145-1). Note the Ersatz wooden grips.***
Photographs by A.Vanderlinden, courtesy Ron Pinkenburg collection

Occupation Production: First alphabet prefix VIS N° H0448

(Fig. 171-1) First Alphabet Series
Serial Number: ***H0448***
WaA Markings: ***WaA77 , E/623***
Slotted for Shoulder-Stock: ***No***
Notes: ***Rougher finish but with early markings. As wartime production demands increased, the quality of fit and finish decreased.***
Photographs by A. Vanderlinden, courtesy Waclaw Ustupski collection

Occupation Production: First alphabet prefix VIS N° J3557

(Fig. 172-1) First Alphabet Series
Serial Number: **J3557**
WaA Markings: **WaA77 , E/623**
Slotted for Shoulder-Stock: **No**
Notes: **The decline in finish quality is becoming more visible on this pistol. The pistol nonetheless remains a quality arm.**
Photographs by A.Vanderlinden, courtesy Waclaw Ustupski collection

Occupation Production: First alphabet prefix VIS N° K3228

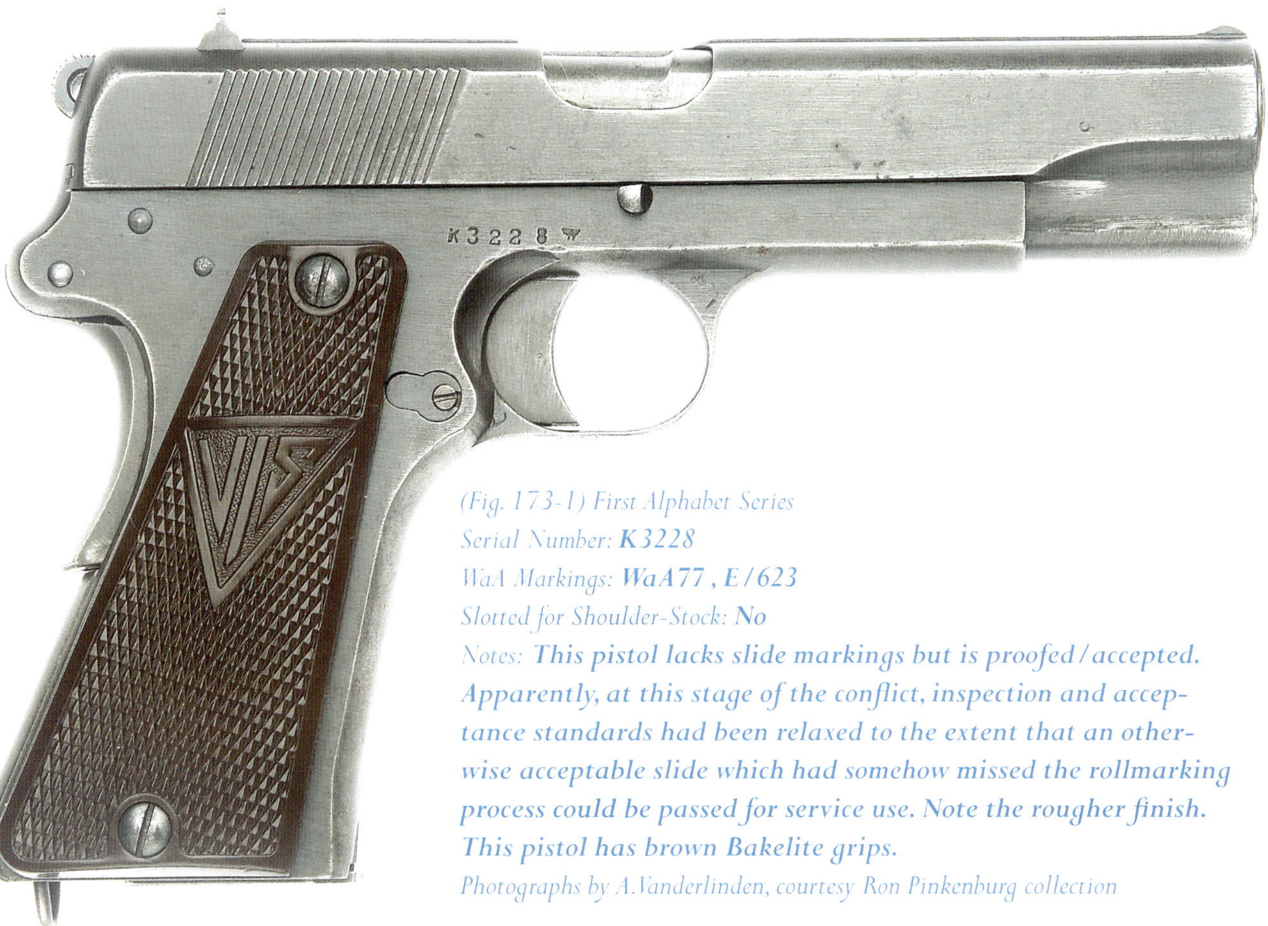

(Fig. 173-1) First Alphabet Series
Serial Number: ***K3228***
WaA Markings: ***WaA77 , E/623***
Slotted for Shoulder-Stock: ***No***
Notes: ***This pistol lacks slide markings but is proofed/accepted. Apparently, at this stage of the conflict, inspection and acceptance standards had been relaxed to the extent that an otherwise acceptable slide which had somehow missed the rollmarking process could be passed for service use. Note the rougher finish. This pistol has brown Bakelite grips.***
Photographs by A.Vanderlinden, courtesy Ron Pinkenburg collection

Occupation Production: First alphabet prefix VIS N° K4730

(Fig. 174-1) First Alphabet Series
Serial Number: **K4730**
WaA Markings: **WaA77 , E/623**
Slotted for Shoulder-Stock: **No**
Notes: ***Note the wartime simplification: A lack of clearance cuts on the pistol's frame behind the trigger. This pistol also has a Waffenamt 77 on the right side of the frame following the serial number.***
Photographs by A.Vanderlinden, courtesy Waclaw Ustupski collection

Occupation Production: First alphabet prefix VIS N° K5906

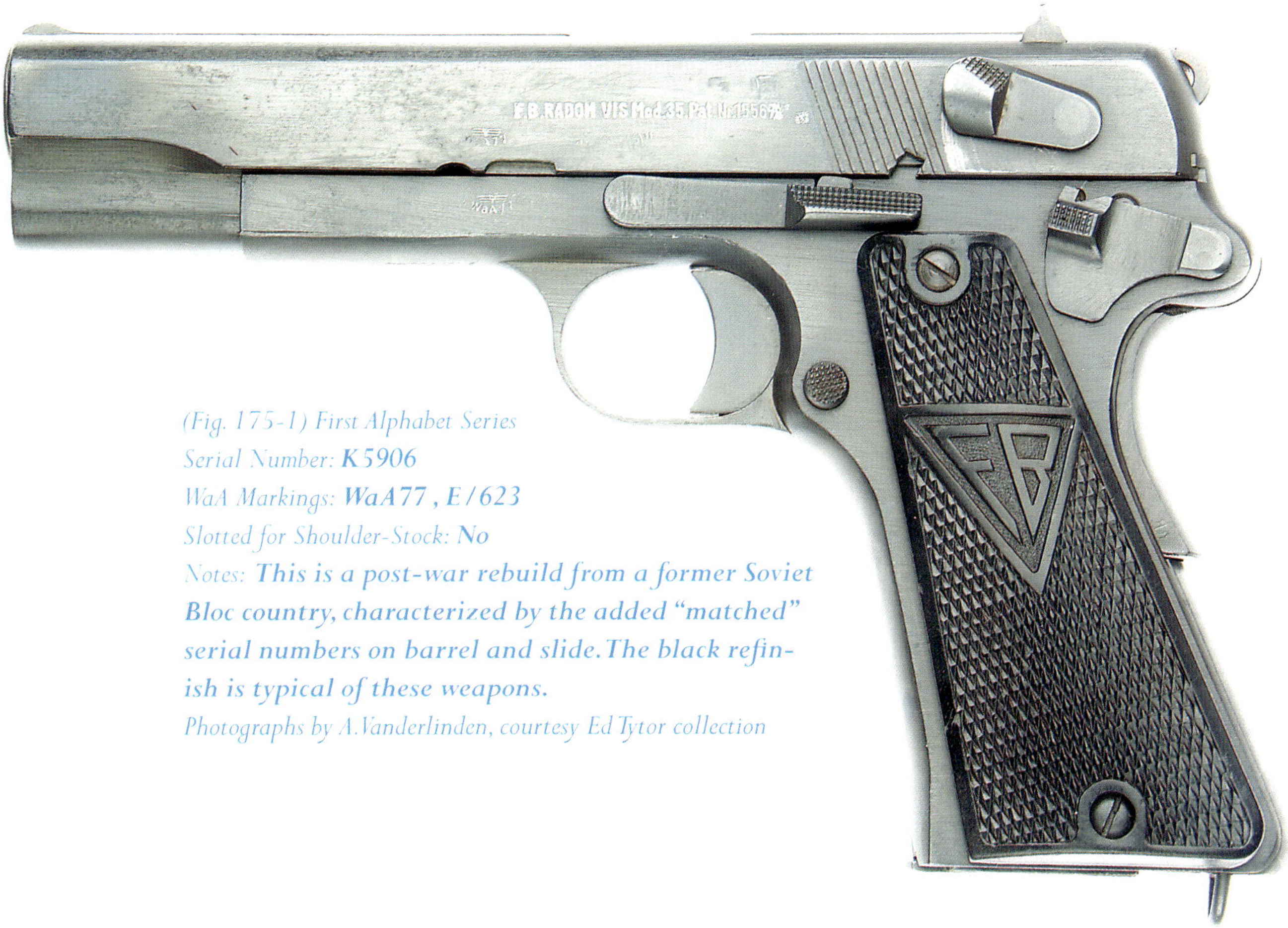

(Fig. 175-1) First Alphabet Series
Serial Number: ***K5906***
WaA Markings: ***WaA77 , E/623***
Slotted for Shoulder-Stock: ***No***
Notes: ***This is a post-war rebuild from a former Soviet Bloc country, characterized by the added "matched" serial numbers on barrel and slide. The black refinish is typical of these weapons.***
Photographs by A. Vanderlinden, courtesy Ed Tytor collection

Occupation Production: First alphabet prefix VIS N° L0212

(Fig. 176-1) First Alphabet Series
Serial Number: ***L0212***
WaA Markings: ***Eagle 77, WaA77, E/623***
Slotted for Shoulder-Stock: ***No***
Notes: ***Note the transition to the Eagle 77 marking instead of the WaA77 marking on the slide.***
Photographs by A.Vanderlinden, courtesy Waclaw Ustupski collection

Occupation Production: First alphabet prefix VIS N° M0079

(Fig. 177-1) First Alphabet Series
Serial Number: ***M0079***
WaA Markings: ***WaA77, E/623***
Slotted for Shoulder-Stock: ***No***
Notes: ***Note the WaA77 marking on frame and slide and the double-struck P35(p) marking on the slide.***
Photographs by A.Vanderlinden, courtesy Waclaw Ustupski collection

Occupation Production: First alphabet prefix VIS N° M2983

(Fig. 178-1) First Alphabet Series
Serial Number: ***M2983***
WaA Markings: ***WaA77, E/623***
Slotted for Shoulder-Stock: ***No***
Notes: ***This example was arsenal rebuilt after World War II. The Waffenamt 77 marking is still visible, but note the renumbering on barrel and slide. A "Century Arms" import mark is located on the bottom of the triggerguard.***
Photographs by A.Vanderlinden, courtesy Waclaw Ustupski collection

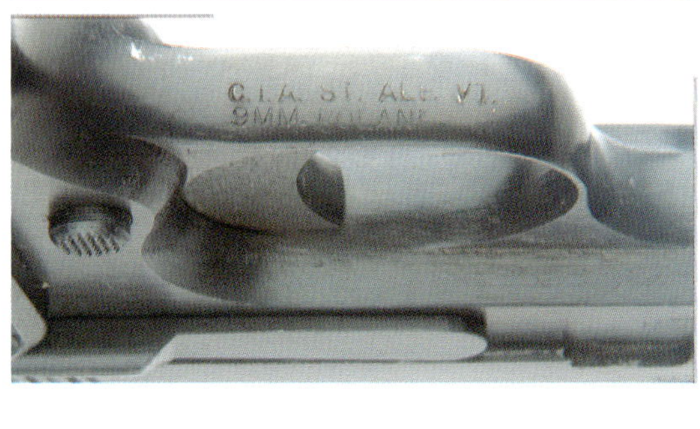

Occupation Production: First alphabet prefix VIS N° M3292

(Fig. 179-1) First Alphabet Series
Serial Number: ***M3292***
WaA Markings: ***WaA77, E/623***
Slotted for Shoulder-Stock: ***No***
Notes: ***This pistol was never finished and left in-the-white. It was assembled with select blued parts and fully proofed and accepted.***
Photographs courtesy of Waclaw Ustupski, from the Steve Vandtke collection

Occupation Production: First alphabet prefix VIS N° N6003

(Fig. 180-1) First Alphabet Series
Serial Number: **N6003**
WaA Markings: **E/77, E/623**
Slotted for Shoulder-Stock: **No**
Notes: ***This pistol no longer has the P35(p) marking on its slide. Note also the Eagle 77 (Radom) markings on slide and frame.***
Photographs by A.Vanderlinden, courtesy Waclaw Ustupski collection

Occupation Production: First alphabet prefix VIS N° N6139

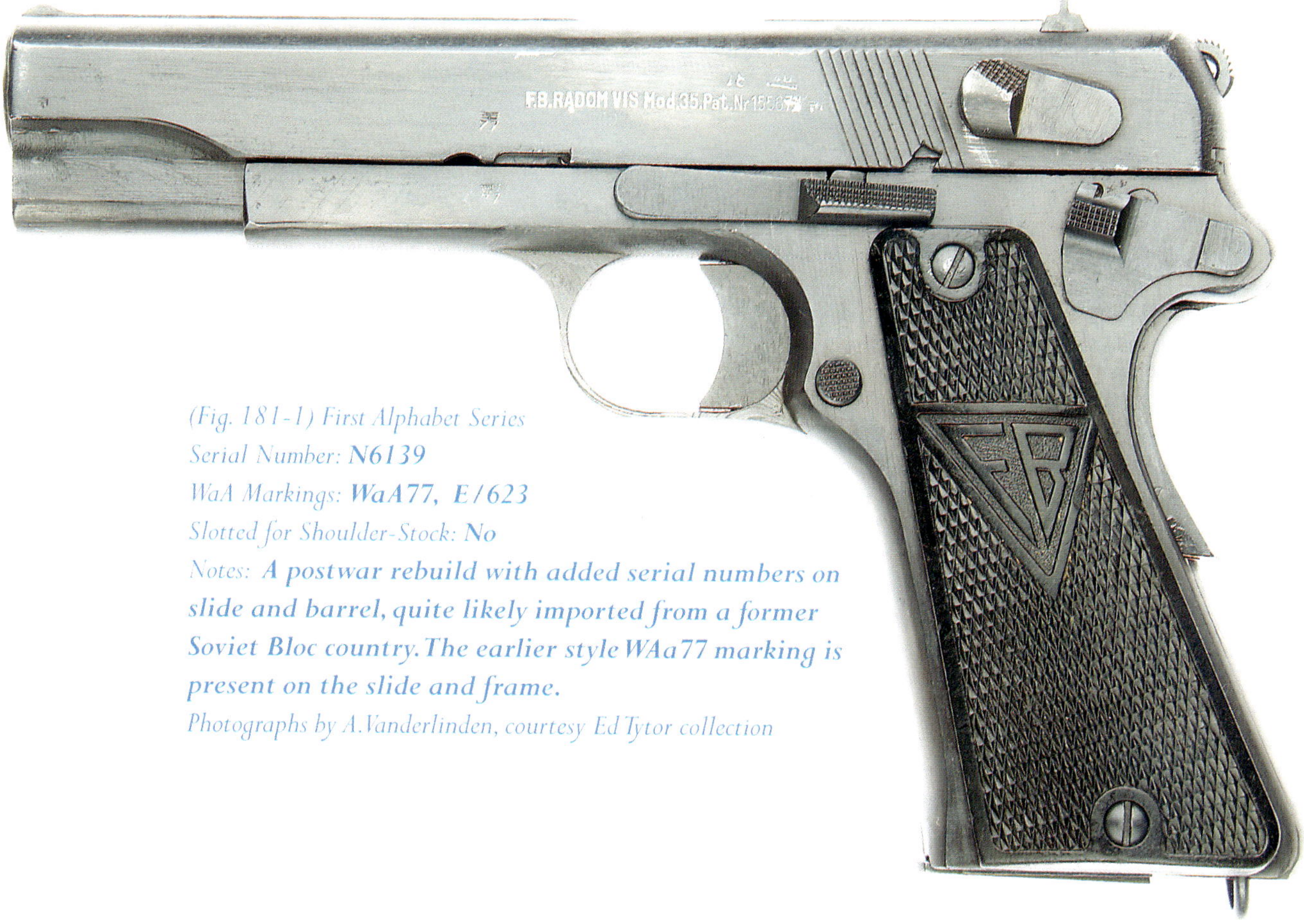

(Fig. 181-1) First Alphabet Series
Serial Number: ***N6139***
WaA Markings: ***WaA77, E/623***
Slotted for Shoulder-Stock: ***No***
Notes: ***A postwar rebuild with added serial numbers on slide and barrel, quite likely imported from a former Soviet Bloc country. The earlier style WAa77 marking is present on the slide and frame.***
Photographs by A. Vanderlinden, courtesy Ed Tytor collection

Occupation Production: First alphabet prefix VIS Nº P3620

(Fig. 182-1) First Alphabet Series
Serial Number: **P3620**
WaA Markings: **E/77, E/623**
Slotted for Shoulder-Stock: **No**
Notes: **This pistol shows the later E/77 marking. The pistol still has small grip screws and a milled trigger.**
Photographs by A.Vanderlinden, courtesy Waclaw Ustupski collection

Occupation Production: First alphabet prefix VIS N° R3421

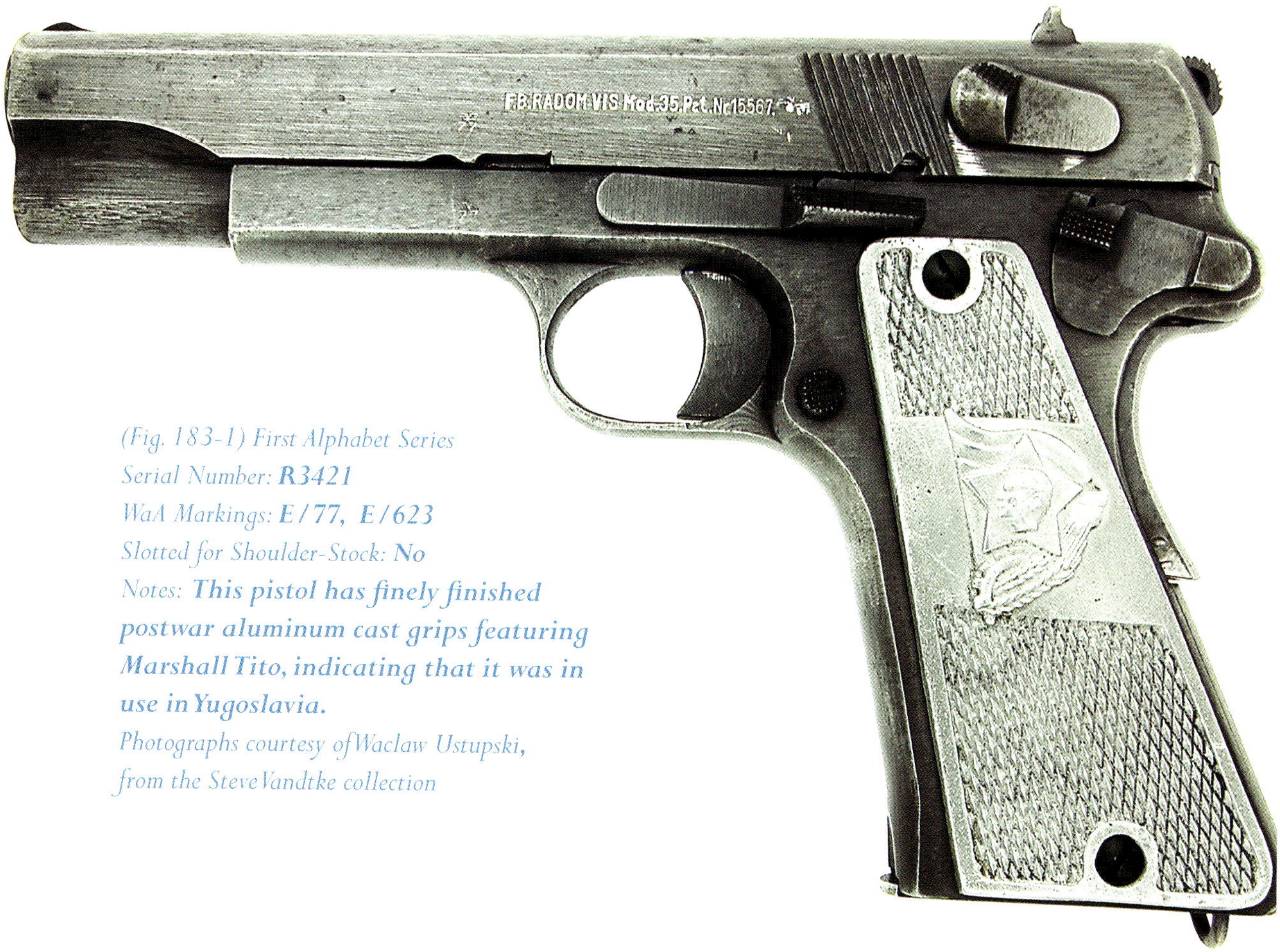

(Fig. 183-1) First Alphabet Series
Serial Number: ***R3421***
WaA Markings: ***E/77, E/623***
Slotted for Shoulder-Stock: ***No***
Notes: ***This pistol has finely finished postwar aluminum cast grips featuring Marshall Tito, indicating that it was in use in Yugoslavia.***
Photographs courtesy of Waclaw Ustupski, from the Steve Vandtke collection

Occupation Production: First alphabet prefix VIS Nº R3691

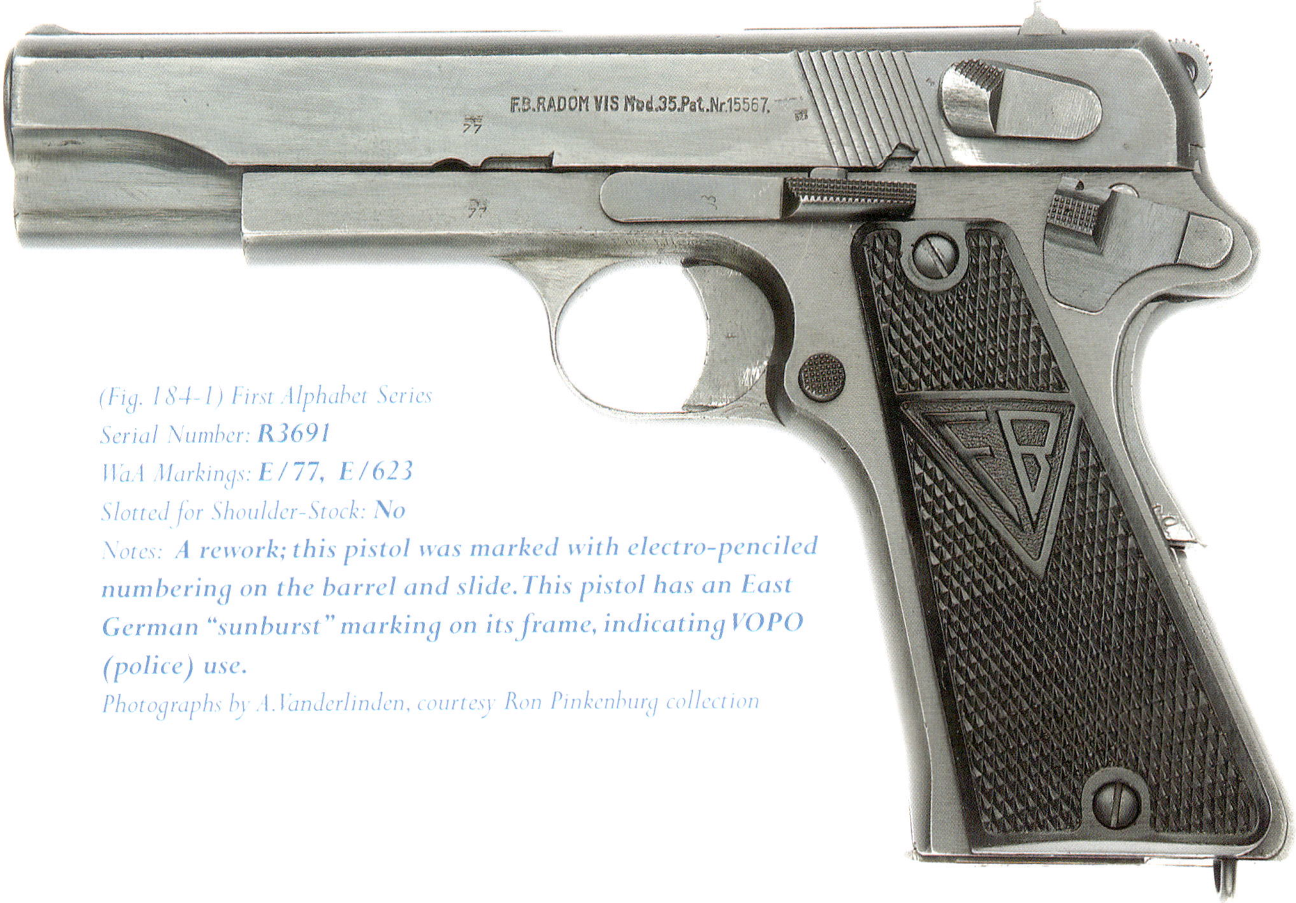

(Fig. 184-1) First Alphabet Series
Serial Number: ***R3691***
WaA Markings: ***E/77, E/623***
Slotted for Shoulder-Stock: ***No***
Notes: ***A rework; this pistol was marked with electro-penciled numbering on the barrel and slide. This pistol has an East German "sunburst" marking on its frame, indicating VOPO (police) use.***
Photographs by A. Vanderlinden, courtesy Ron Pinkenburg collection

Occupation Production: First alphabet prefix VIS N° R5682

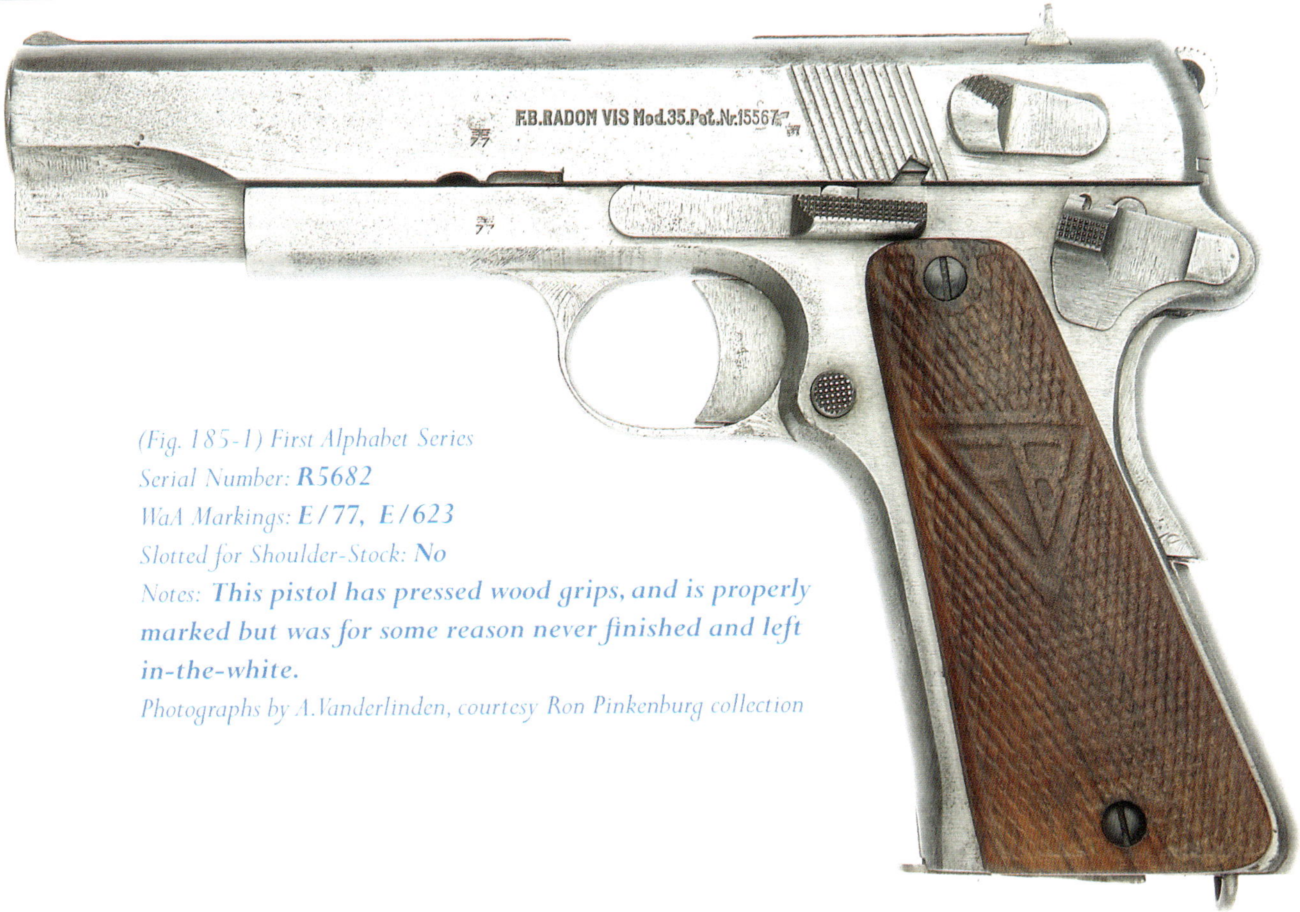

(Fig. 185-1) First Alphabet Series
Serial Number: **R5682**
WaA Markings: **E/77, E/623**
Slotted for Shoulder-Stock: **No**
Notes: ***This pistol has pressed wood grips, and is properly marked but was for some reason never finished and left in-the-white.***
Photographs by A.Vanderlinden, courtesy Ron Pinkenburg collection

Occupation Production: First alphabet prefix VIS N° R5724

(Fig. 186-1) First Alphabet Series
Serial Number: ***R5724***
WaA Markings: ***E/77, E/623***
Slotted for Shoulder-Stock: ***No***
Notes: ***Around this serial number range, there is a change in the number of grasping grooves on the left side of the slide, from 10 to 11. There is no change on the right side. Note the simplified markings and the visibly cruder finish.***
Photographs by A.Vanderlinden, courtesy Waclaw Ustupski collection

Occupation Production: First alphabet prefix VIS N° S2421

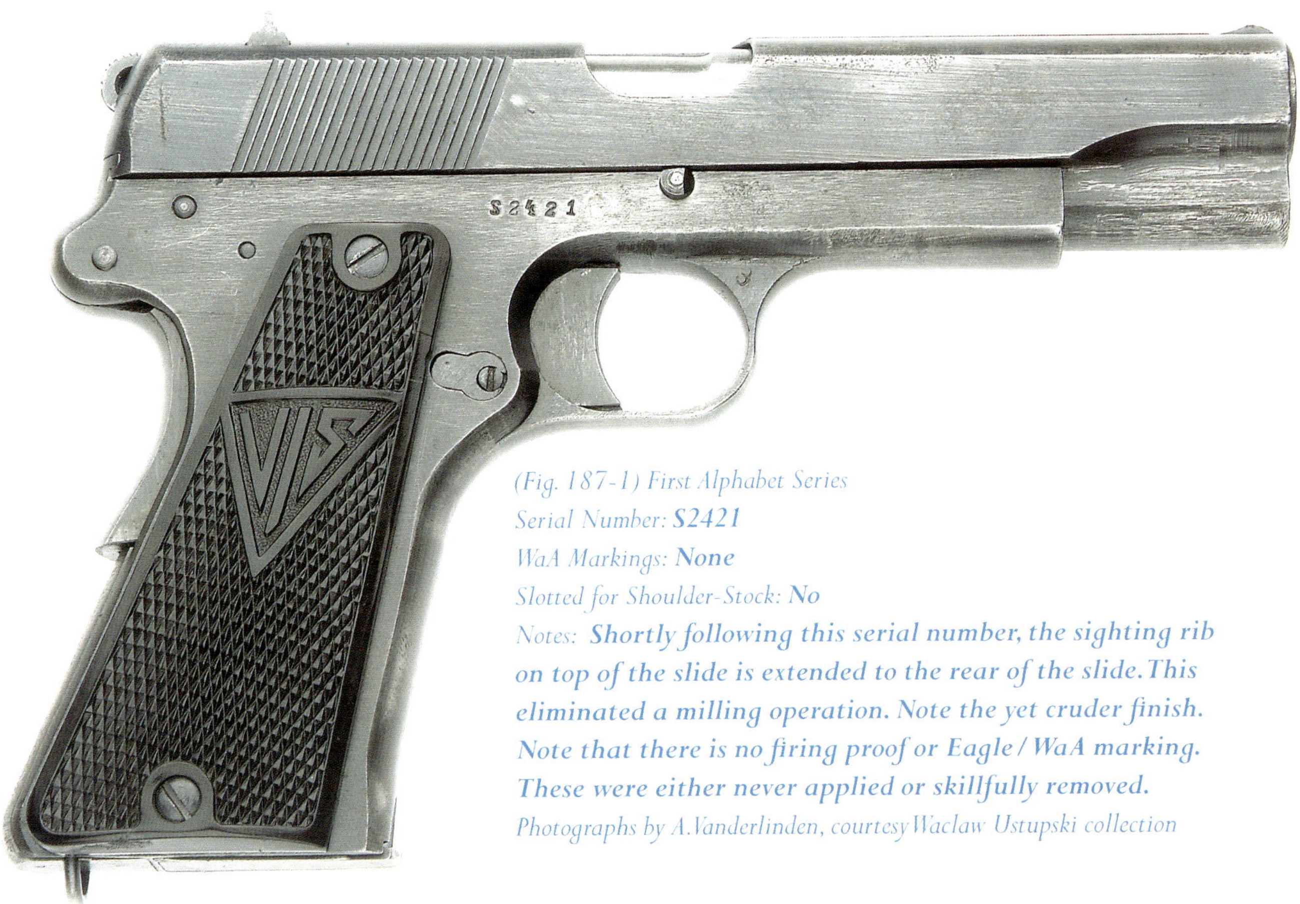

(Fig. 187-1) First Alphabet Series

Serial Number: ***S2421***

WaA Markings: ***None***

Slotted for Shoulder-Stock: ***No***

Notes: ***Shortly following this serial number, the sighting rib on top of the slide is extended to the rear of the slide. This eliminated a milling operation. Note the yet cruder finish. Note that there is no firing proof or Eagle / WaA marking. These were either never applied or skillfully removed.***

Photographs by A.Vanderlinden, courtesy Waclaw Ustupski collection

Occupation Production: First alphabet prefix VIS N° T6371

(Fig. 188-1) First Alphabet Series
Serial Number: **T6371**
WaA Markings: **E/77, E/623**
Slotted for Shoulder-Stock: **No**
Notes: ***The sighting rib is no longer milled off behind the rear sight after the mid-S serial number bloc.***
Photographs by A.Vanderlinden, courtesy Waclaw Ustupski collection

Occupation Production: First alphabet prefix VIS N° U1016

(Fig. 189-1) First Alphabet Series
Serial Number: ***U1016***
WaA Markings: ***E/77, E/623***
Slotted for Shoulder-Stock: ***No***
Notes: ***Note the pressed wood grips.***
Photographs by A.Vanderlinden, courtesy Francis Kennedy collection

Occupation Production: First alphabet prefix VIS N° U5247

(Fig. 190-1) First Alphabet Series
Serial Number: ***U5247***
WaA Markings: ***E/77, E/623***
Slotted for Shoulder-Stock: ***No***
Notes: ***The trigger on this pistol is one of the last milled triggers that were assembled.***
Photographs by A.Vanderlinden, courtesy Waclaw Ustupski collection

Occupation Production: First alphabet prefix VIS N° W2721

(Fig. 191-1) First Alphabet Series
Serial Number: ***W2721***
WaA Markings: ***E/77, E/623***
Slotted for Shoulder-Stock: ***No***
Notes: ***Note the introduction of a stamped and fabricated trigger versus the milled trigger as seen on the previous pistol.***
Photographs by A.Vanderlinden, courtesy Waclaw Ustupski collection

Occupation Production: First alphabet prefix VIS Nº Z1035

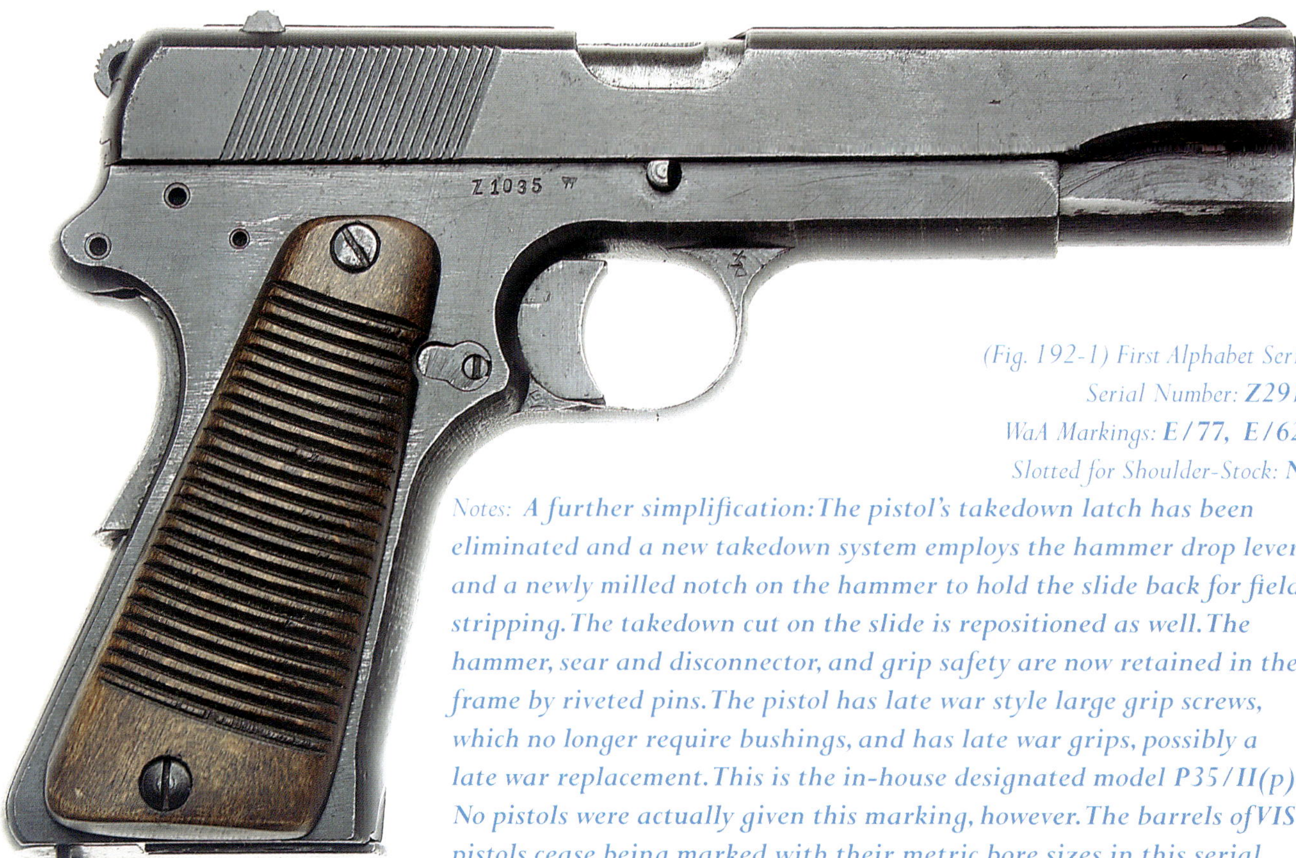

(Fig. 192-1) First Alphabet Series

Serial Number: ***Z2919***

WaA Markings: ***E/77, E/623***

Slotted for Shoulder-Stock: ***No***

Notes: ***A further simplification: The pistol's takedown latch has been eliminated and a new takedown system employs the hammer drop lever and a newly milled notch on the hammer to hold the slide back for field stripping. The takedown cut on the slide is repositioned as well. The hammer, sear and disconnector, and grip safety are now retained in the frame by riveted pins. The pistol has late war style large grip screws, which no longer require bushings, and has late war grips, possibly a late war replacement. This is the in-house designated model P35/II(p). No pistols were actually given this marking, however. The barrels of VIS pistols cease being marked with their metric bore sizes in this serial number range.*** *Photographs by A.Vanderlinden, courtesy Francis Kennedy collection*

Occupation Production: First alphabet prefix VIS N° Z2581

(Fig. 193-1) First Alphabet Series
Serial Number: ***Z2581***
WaA Markings: ***E/77, E/623***
Slotted for Shoulder-Stock: ***No***
Notes: ***This example has the new fabricated trigger and retains the takedown lever and small grip screws with bushings, which are phased out in this serial number range.***
Photographs by A.Vanderlinden, courtesy Waclaw Ustupski collection

Occupation Production: First alphabet prefix VIS N° Z3069

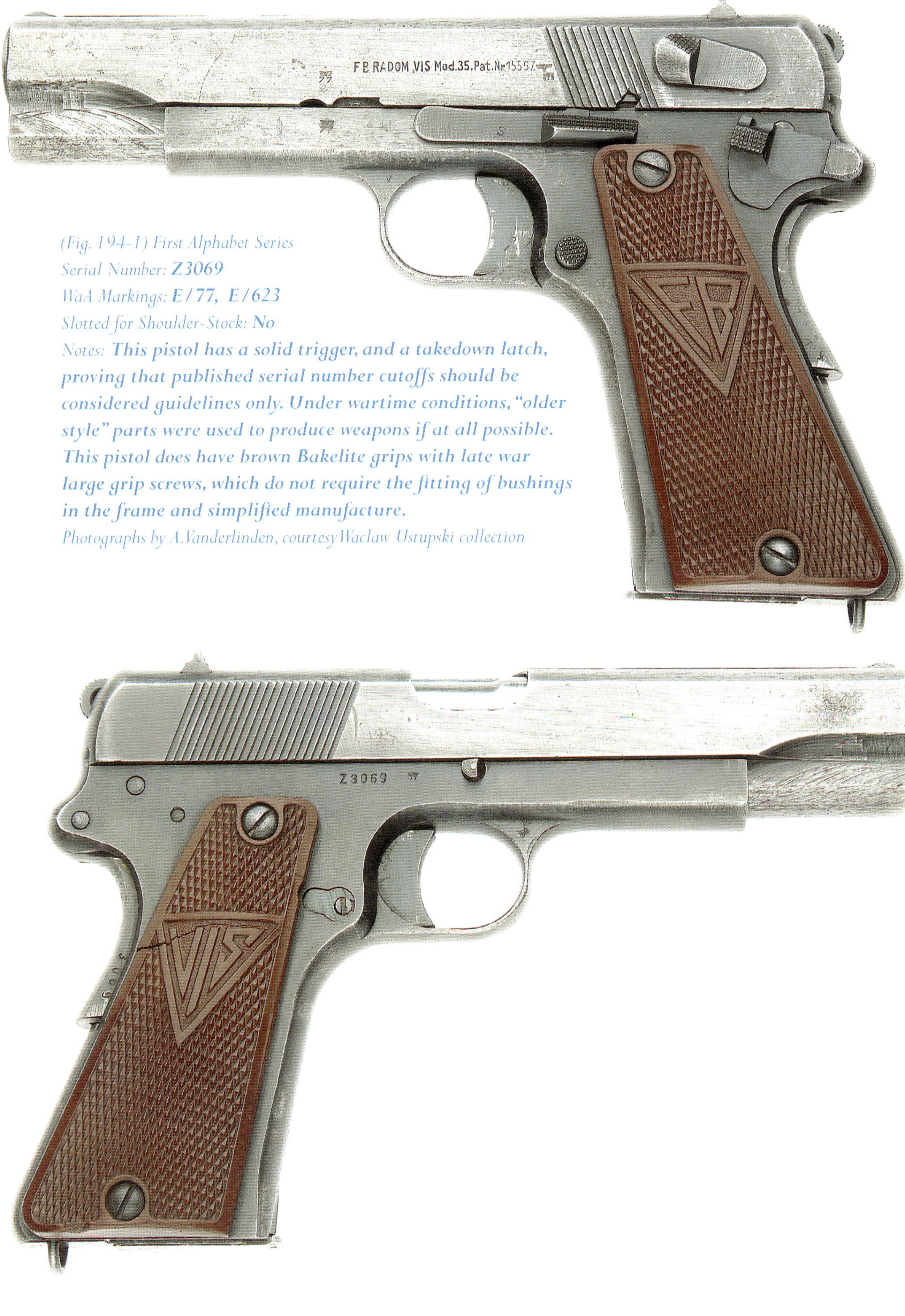

(Fig. 194-1) First Alphabet Series
Serial Number: ***Z3069***
WaA Markings: ***E/77, E/623***
Slotted for Shoulder-Stock: ***No***
Notes: ***This pistol has a solid trigger, and a takedown latch, proving that published serial number cutoffs should be considered guidelines only. Under wartime conditions, "older style" parts were used to produce weapons if at all possible. This pistol does have brown Bakelite grips with late war large grip screws, which do not require the fitting of bushings in the frame and simplified manufacture.***
Photographs by A.Vanderlinden, courtesy Waclaw Ustupski collection

Occupation Production: First alphabet prefix VIS Nº Z9706

(Fig. 195-1) First Alphabet Series
Serial Number: **Z9706**
WaA Markings: **E/77, E/623**
Slotted for Shoulder-Stock: **No**
Notes: ***This pistol has a blued barrel, a machined trigger, brown Bakelite grips and rough machining marks.***
Photographs by Anthony Vanderlinden

Photography Gallery III

Occupation Production at Radom & Steyr/Znaim Second Alphabet Prefix Series

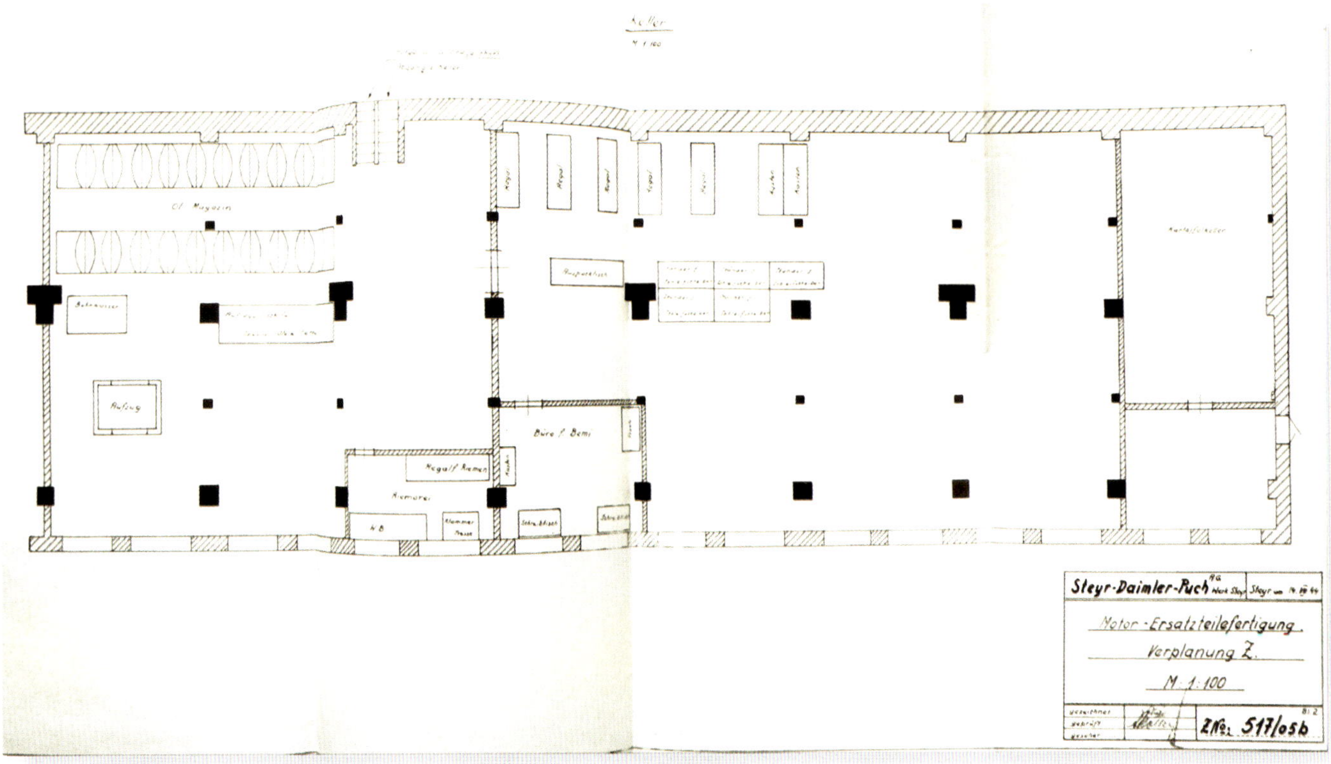

(Fig. 196-1) Early Steyr plans of the basement of the Motor-Ersatzteilefertigung in Znaim where theVIS production was moved to. (CourtesyWaclaw Ustupski)

Occupation Production: Second alphabet prefix VIS N° A5340

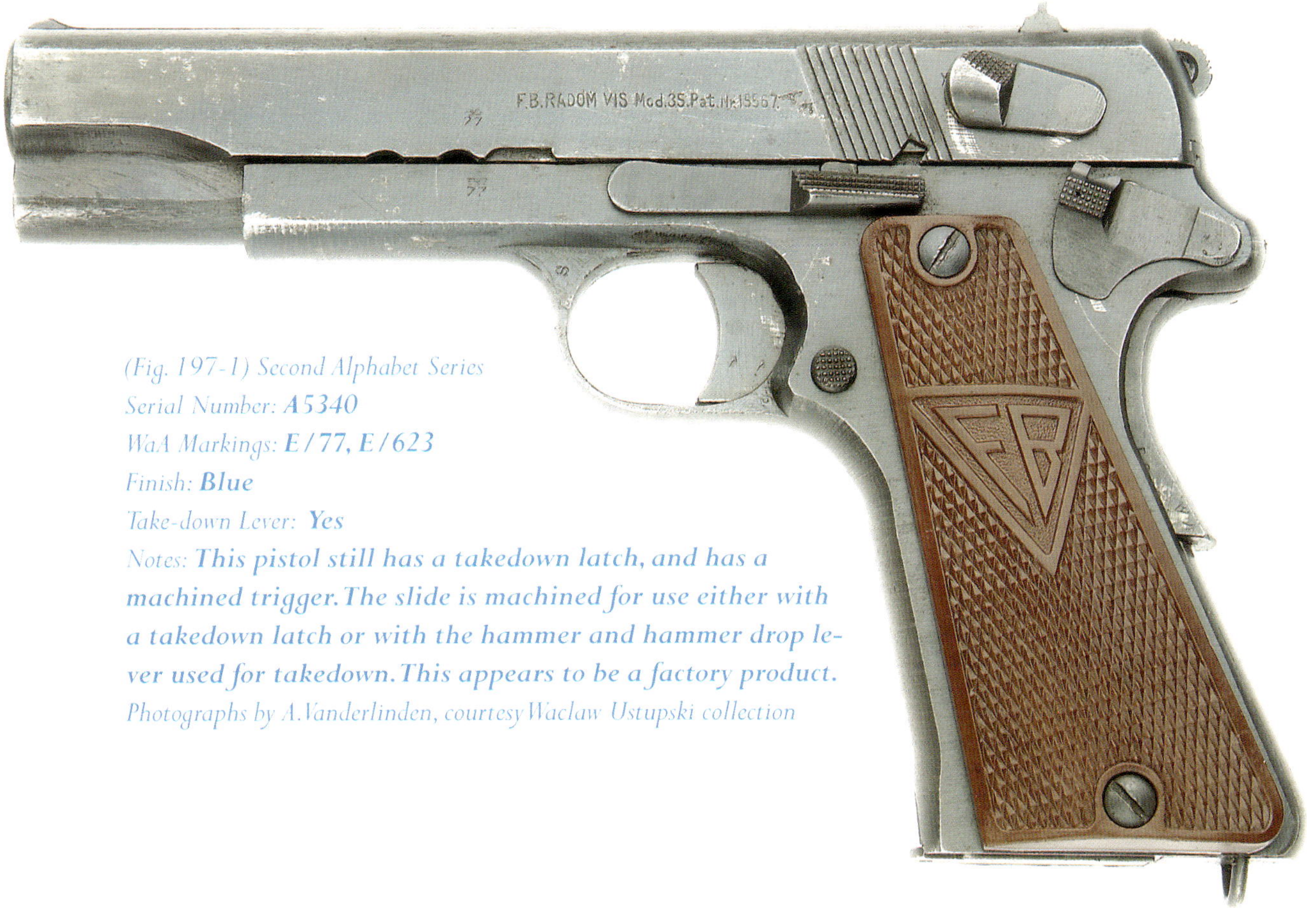

(Fig. 197-1) Second Alphabet Series
Serial Number: ***A5340***
WaA Markings: ***E/77, E/623***
Finish: ***Blue***
Take-down Lever: ***Yes***
Notes: ***This pistol still has a takedown latch, and has a machined trigger. The slide is machined for use either with a takedown latch or with the hammer and hammer drop lever used for takedown. This appears to be a factory product.***
Photographs by A.Vanderlinden, courtesy Waclaw Ustupski collection

Occupation Production: Second alphabet prefix VIS N° A6836

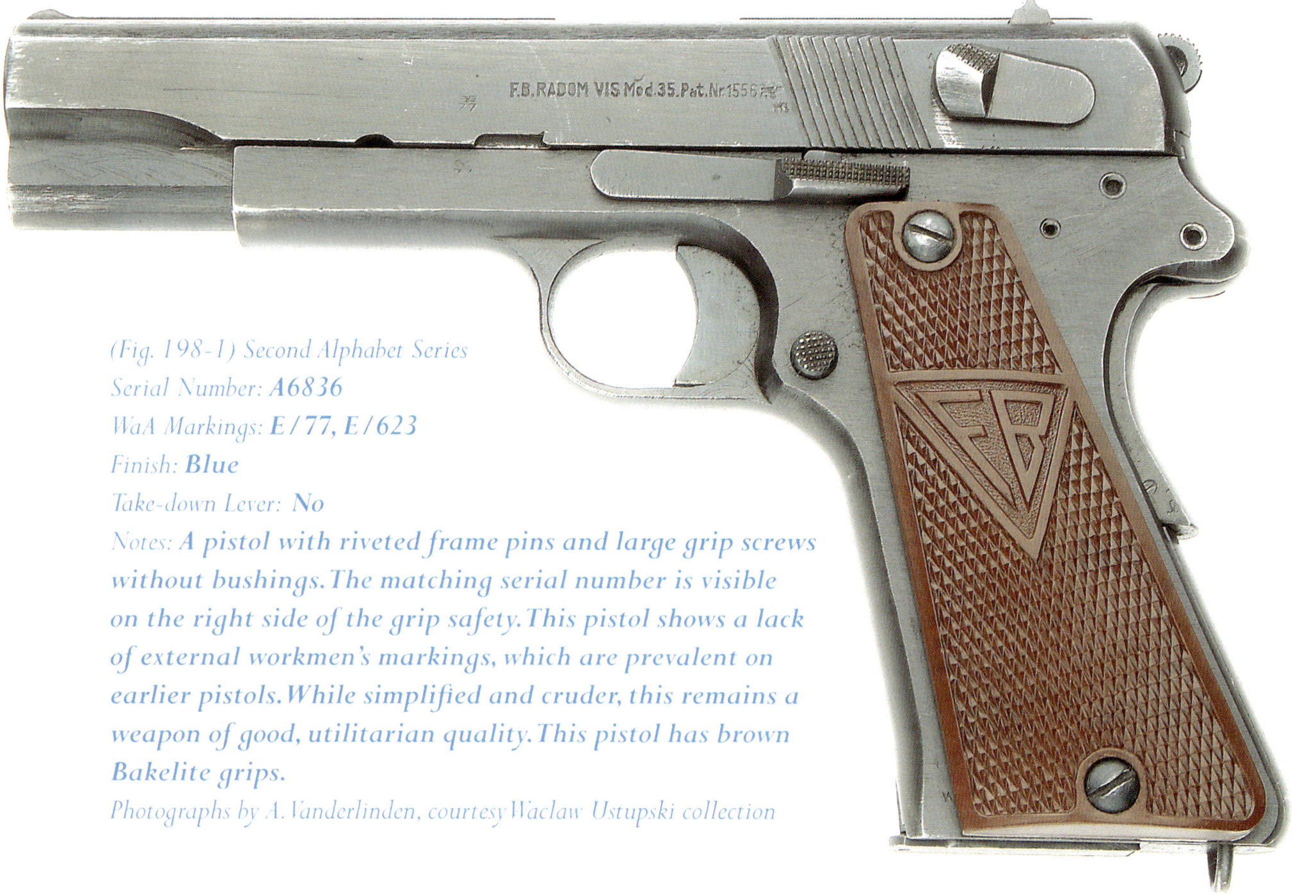

(Fig. 198-1) Second Alphabet Series
Serial Number: ***A6836***
WaA Markings: ***E/77, E/623***
Finish: ***Blue***
Take-down Lever: ***No***
Notes: ***A pistol with riveted frame pins and large grip screws without bushings. The matching serial number is visible on the right side of the grip safety. This pistol shows a lack of external workmen's markings, which are prevalent on earlier pistols. While simplified and cruder, this remains a weapon of good, utilitarian quality. This pistol has brown Bakelite grips.***
Photographs by A. Vanderlinden, courtesy Waclaw Ustupski collection

Occupation Production: Second alphabet prefix VIS Nº A9225

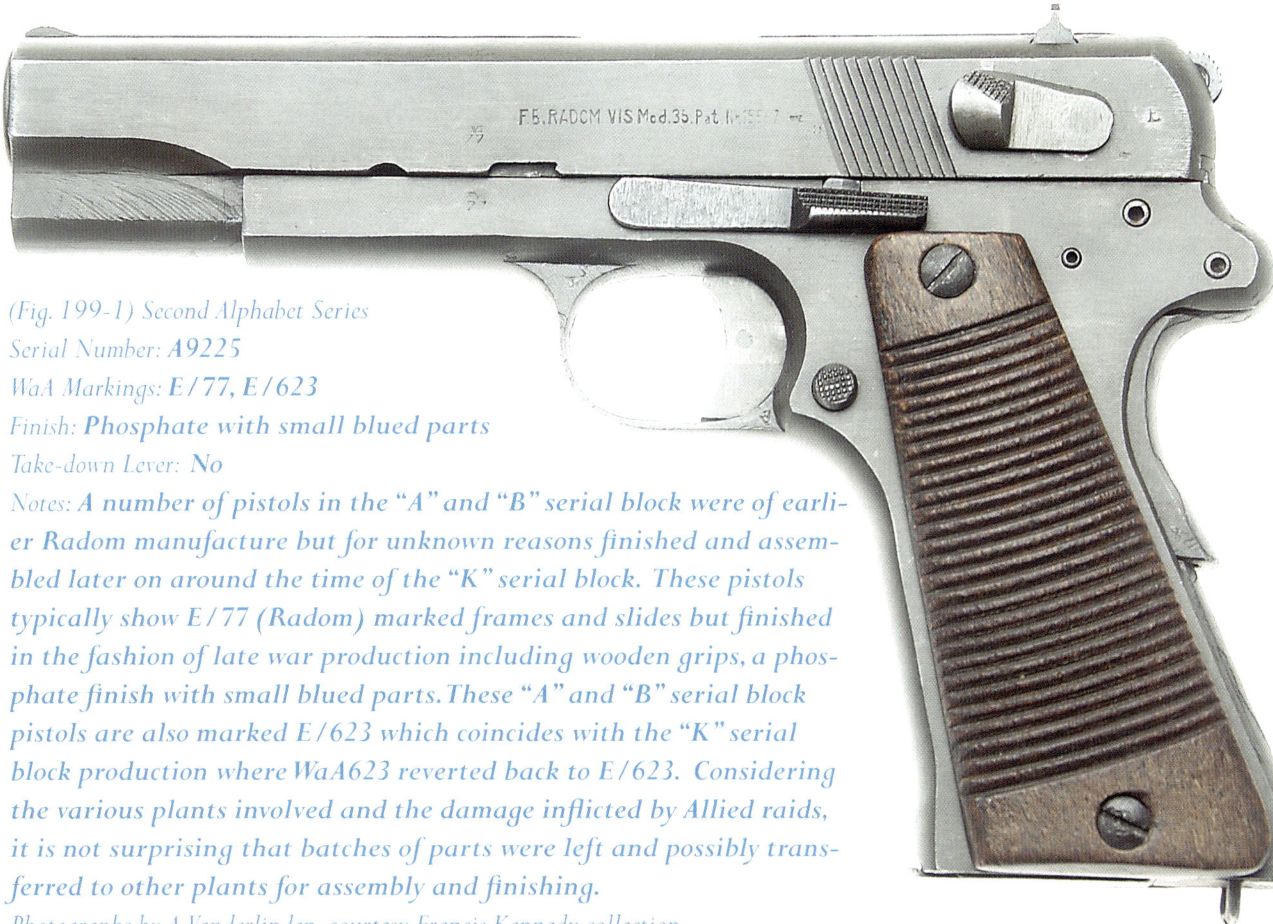

(Fig. 199-1) Second Alphabet Series
Serial Number: ***A9225***
WaA Markings: ***E/77, E/623***
Finish: ***Phosphate with small blued parts***
Take-down Lever: ***No***
Notes: ***A number of pistols in the "A" and "B" serial block were of earlier Radom manufacture but for unknown reasons finished and assembled later on around the time of the "K" serial block. These pistols typically show E/77 (Radom) marked frames and slides but finished in the fashion of late war production including wooden grips, a phosphate finish with small blued parts. These "A" and "B" serial block pistols are also marked E/623 which coincides with the "K" serial block production where WaA623 reverted back to E/623. Considering the various plants involved and the damage inflicted by Allied raids, it is not surprising that batches of parts were left and possibly transferred to other plants for assembly and finishing.***
Photographs by A.Vanderlinden, courtesy Francis Kennedy collection

Occupation Production: Second alphabet prefix VIS N° B4041

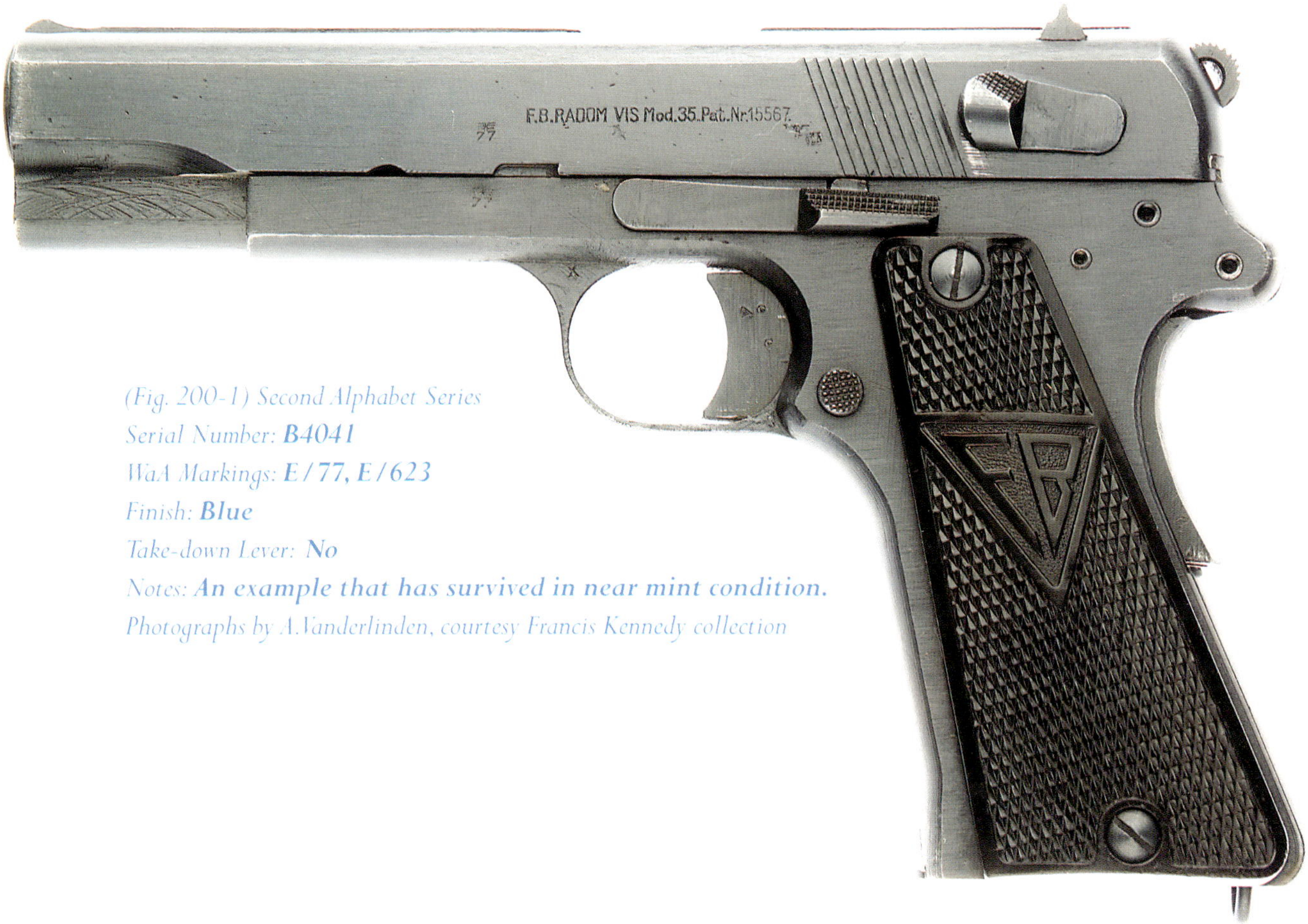

(Fig. 200-1) Second Alphabet Series
Serial Number: **B4041**
WaA Markings: **E/77, E/623**
Finish: **Blue**
Take-down Lever: **No**
Notes: ***An example that has survived in near mint condition.***
Photographs by A.Vanderlinden, courtesy Francis Kennedy collection

Occupation Production: Second alphabet prefix VIS N° B9493

(Fig. 201-1) Second Alphabet Series
Serial Number: **B9493**
WaA Markings: **E/77, WaA623**
Finish: **Blue**
Take-down Lever: **No**
Notes: ***A pistol with riveted frame pins and large grip screws without bushings. As with previous pistols, this weapon still shows good utilitarian quality. External machining is still good with the exception of visible roughness on the lower front of the slide.***
Photographs by A.Vanderlinden, courtesy Waclaw Ustupski collection

Occupation Production: Second alphabet prefix VIS N° C2438

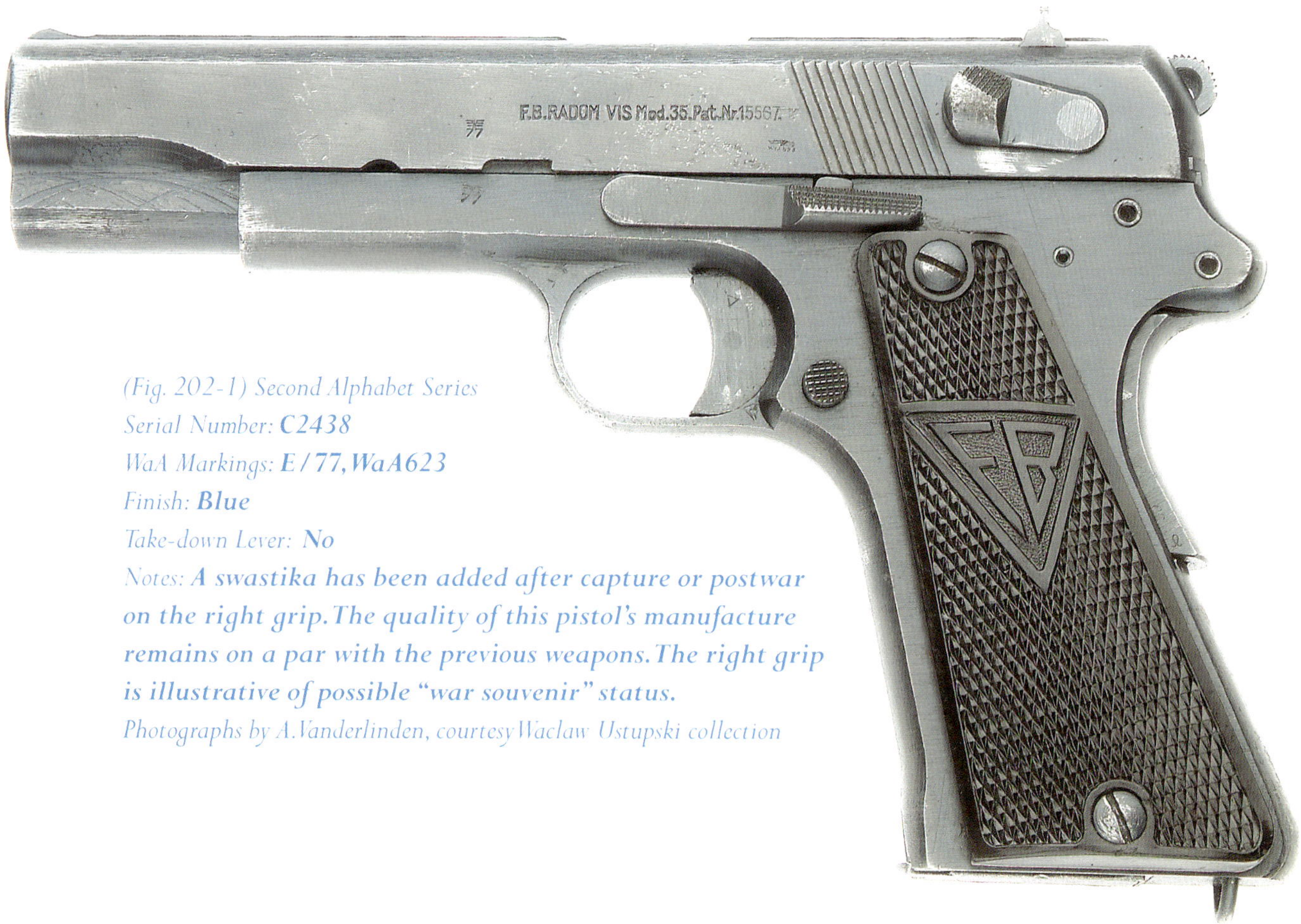

(Fig. 202-1) Second Alphabet Series
Serial Number: ***C2438***
WaA Markings: ***E/77,WaA623***
Finish: ***Blue***
Take-down Lever: ***No***
Notes: ***A swastika has been added after capture or postwar on the right grip. The quality of this pistol's manufacture remains on a par with the previous weapons. The right grip is illustrative of possible "war souvenir" status.***
Photographs by A.Vanderlinden, courtesy Waclaw Ustupski collection

Occupation Production: Second alphabet prefix VIS Nº D6773

(Fig. 203-1) Second Alphabet Series

Serial Number: ***D6773***

WaA Markings: ***E/77, WaA623***

Finish: ***Blue***

Take-down Lever: ***No***

Notes: ***The slide markings on this pistol are quite thin, though the machining of the lower front portion of its slide is visibly much smoother than that on the previously illustrated pistols. Note also the "beta" marking on the slide latch. This pistol has brown Bakelite grips.***

Photographs by A. Vanderlinden, courtesy Waclaw Ustupski collection

Occupation Production: Second alphabet prefix VIS N° E4885

(Fig. 204-1) Second Alphabet Series
Serial Number: **E4885**
WaA Markings: **E/77,WaA623**
Finish: **Blue**
Take-down Lever: **No**
Notes: **In contrast with the previous pistol, this weapon's slide markings are deeply applied, and external machining marks are once more in evidence.**
Photographs by A.Vanderlinden, courtesy Waclaw Ustupski collection

Occupation Production: Second alphabet prefix VIS N° E4934

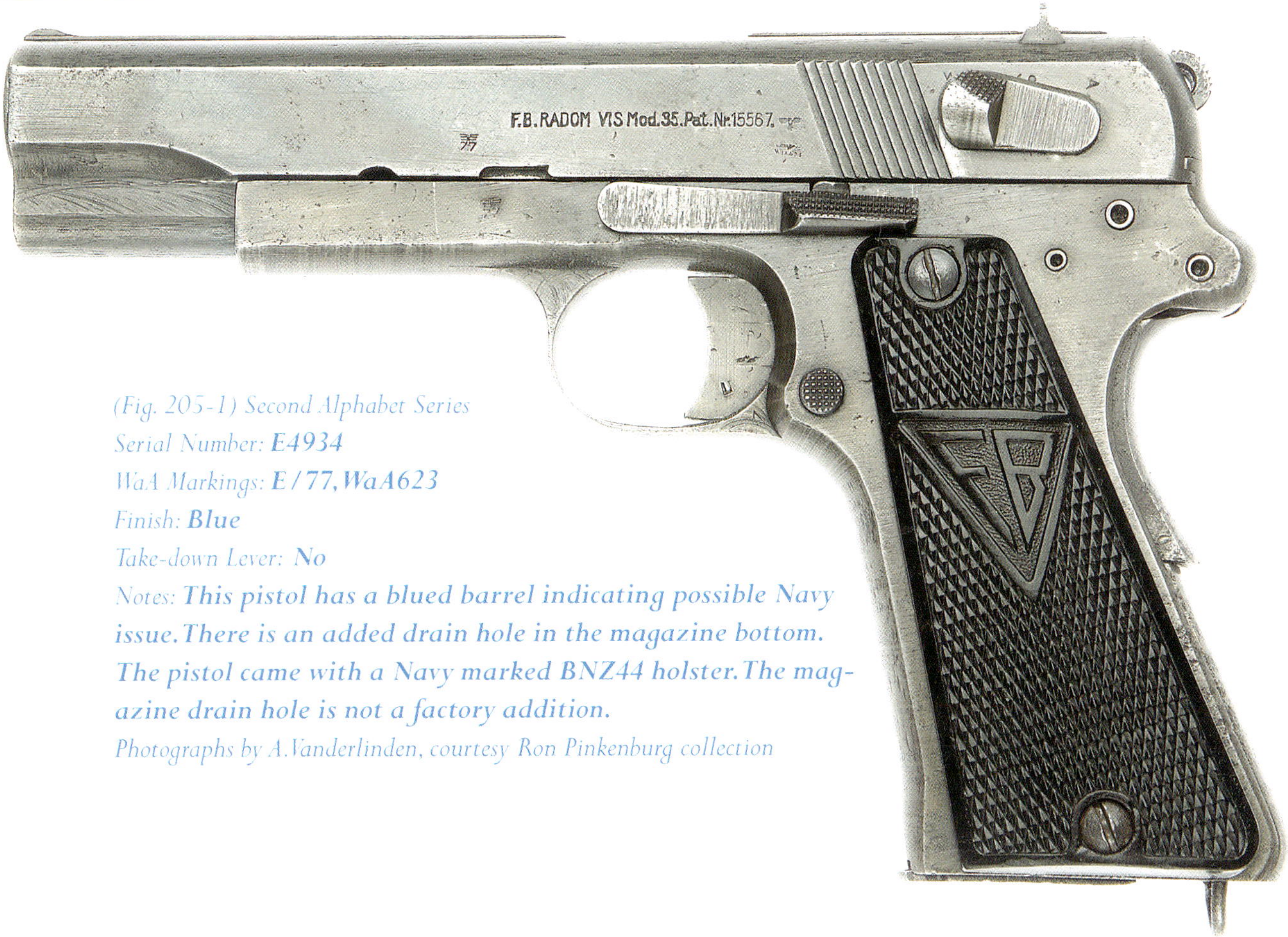

(Fig. 205-1) Second Alphabet Series
Serial Number: ***E4934***
WaA Markings: ***E/77, WaA623***
Finish: ***Blue***
Take-down Lever: ***No***
Notes: ***This pistol has a blued barrel indicating possible Navy issue. There is an added drain hole in the magazine bottom. The pistol came with a Navy marked BNZ44 holster. The magazine drain hole is not a factory addition.***
Photographs by A. Vanderlinden, courtesy Ron Pinkenburg collection

Occupation Production: Second alphabet prefix VIS Nº E4934 (continued)

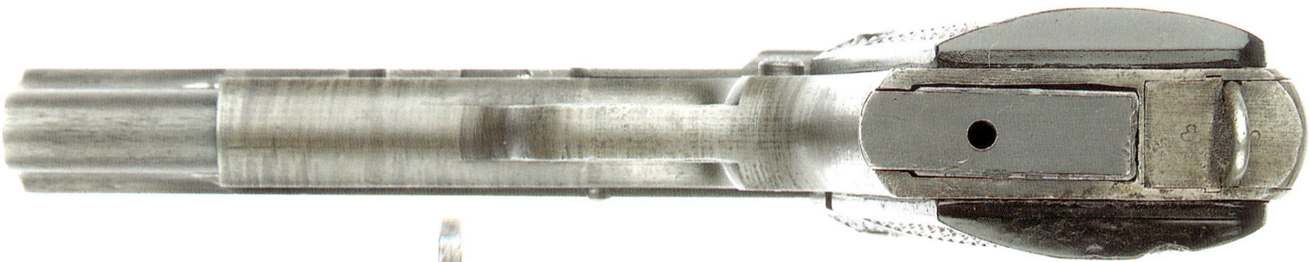

(Fig. 206-1) Notes: ***Note the added drain hole in the magazine floorplate***

(Fig. 206-2) Notes: ***A view of the slide's inventory number.***

(Fig. 206-3) Notes: ***Views of the Navy marked holster and its markings. Holsters with such markings can easily be counterfitted, and caution is advised.***

Occupation Production: Second alphabet prefix VIS N° F4374

(Fig. 207-1) Second Alphabet Series
Serial Number: ***F4374***
WaA Markings: ***E/77, WaA623***
Finish: ***Blue***
Take-down Lever: ***No***
Notes: ***This pistol differs but little in fit and finish from the late wartime pistols previously described- it is still a quite serviceable weapon.***
Photographs by A.Vanderlinden, courtesy Waclaw Ustupski collection

Occupation Production: Second alphabet prefix VIS Nº F5451

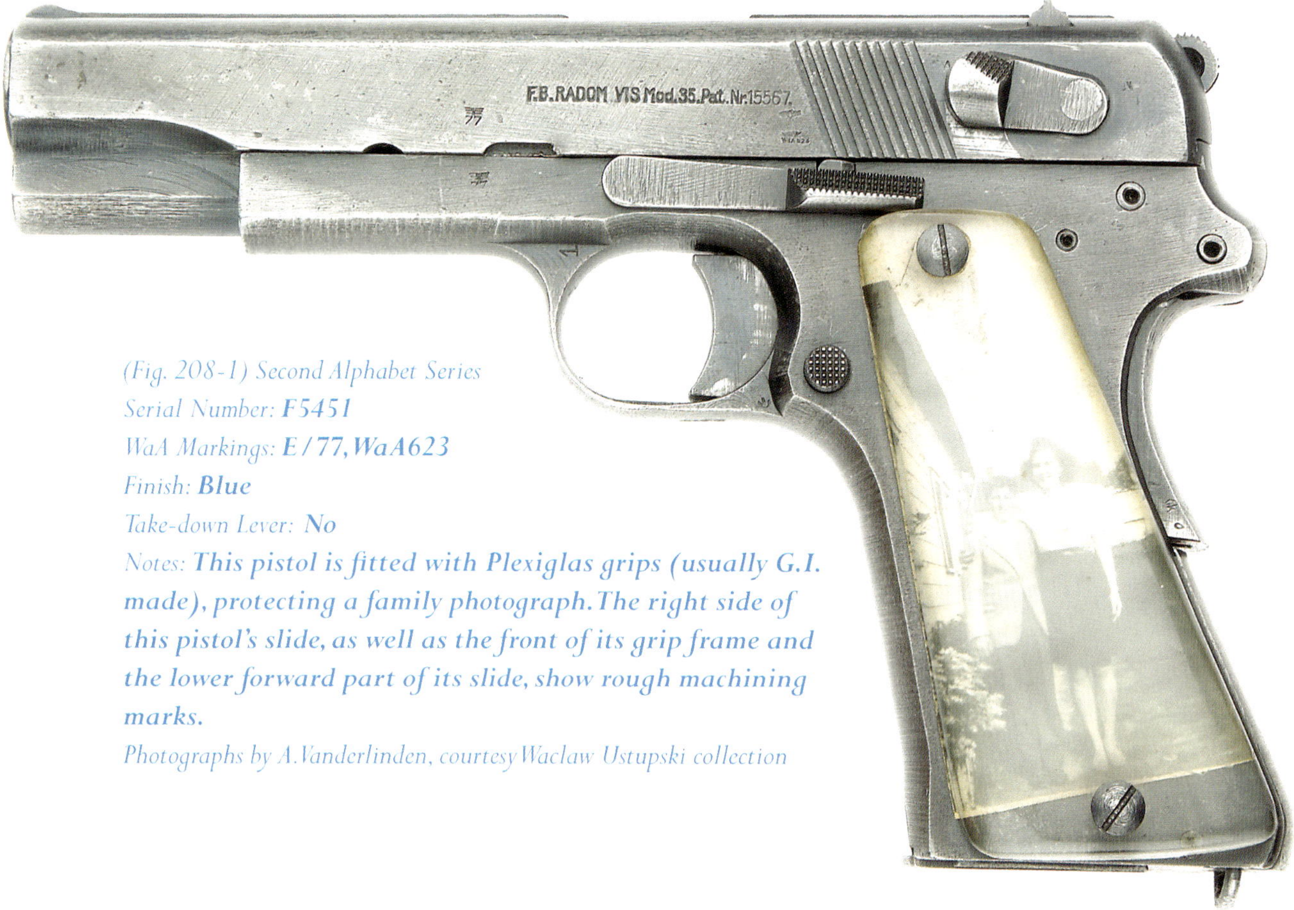

(Fig. 208-1) Second Alphabet Series
Serial Number: ***F5451***
WaA Markings: ***E/77, WaA623***
Finish: ***Blue***
Take-down Lever: ***No***
Notes: ***This pistol is fitted with Plexiglas grips (usually G.I. made), protecting a family photograph. The right side of this pistol's slide, as well as the front of its grip frame and the lower forward part of its slide, show rough machining marks.***
Photographs by A. Vanderlinden, courtesy Waclaw Ustupski collection

Occupation Production: Second alphabet prefix VIS N° G8399

(Fig. 209-1) Second Alphabet Series
Serial Number: ***G8399***
WaA Markings: ***E/77, WaA623***
Finish: ***Blue***
Take-down Lever: ***No***
Notes: ***This pistol shows fewer external machining marks than the previous weapon. Perhaps this is due to a slow-down of production (a form of sabotage) or simply to its assembler's use of the next parts in bin, pile, or basket, fitting some earlier parts to some later pistols.***
Photographs by A.Vanderlinden, courtesy Waclaw Ustupski collection

Occupation Production: Second alphabet prefix VIS N° H4394

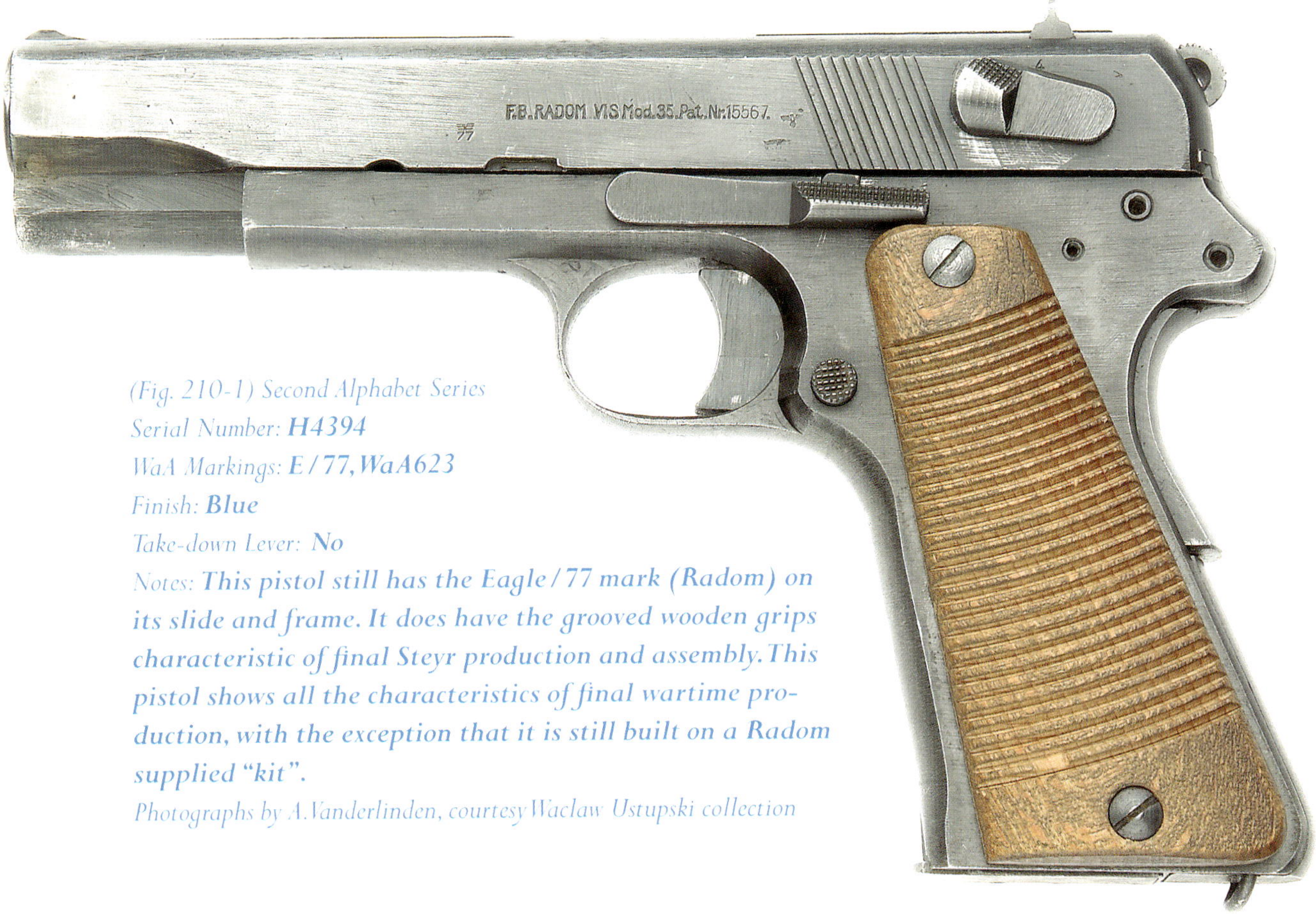

(Fig. 210-1) Second Alphabet Series
Serial Number: **H4394**
WaA Markings: **E/77, WaA623**
Finish: **Blue**
Take-down Lever: **No**
Notes: **This pistol still has the Eagle/77 mark (Radom) on its slide and frame. It does have the grooved wooden grips characteristic of final Steyr production and assembly. This pistol shows all the characteristics of final wartime production, with the exception that it is still built on a Radom supplied "kit".**
Photographs by A. Vanderlinden, courtesy Waclaw Ustupski collection

Occupation Production: Second alphabet prefix VIS N° H9958

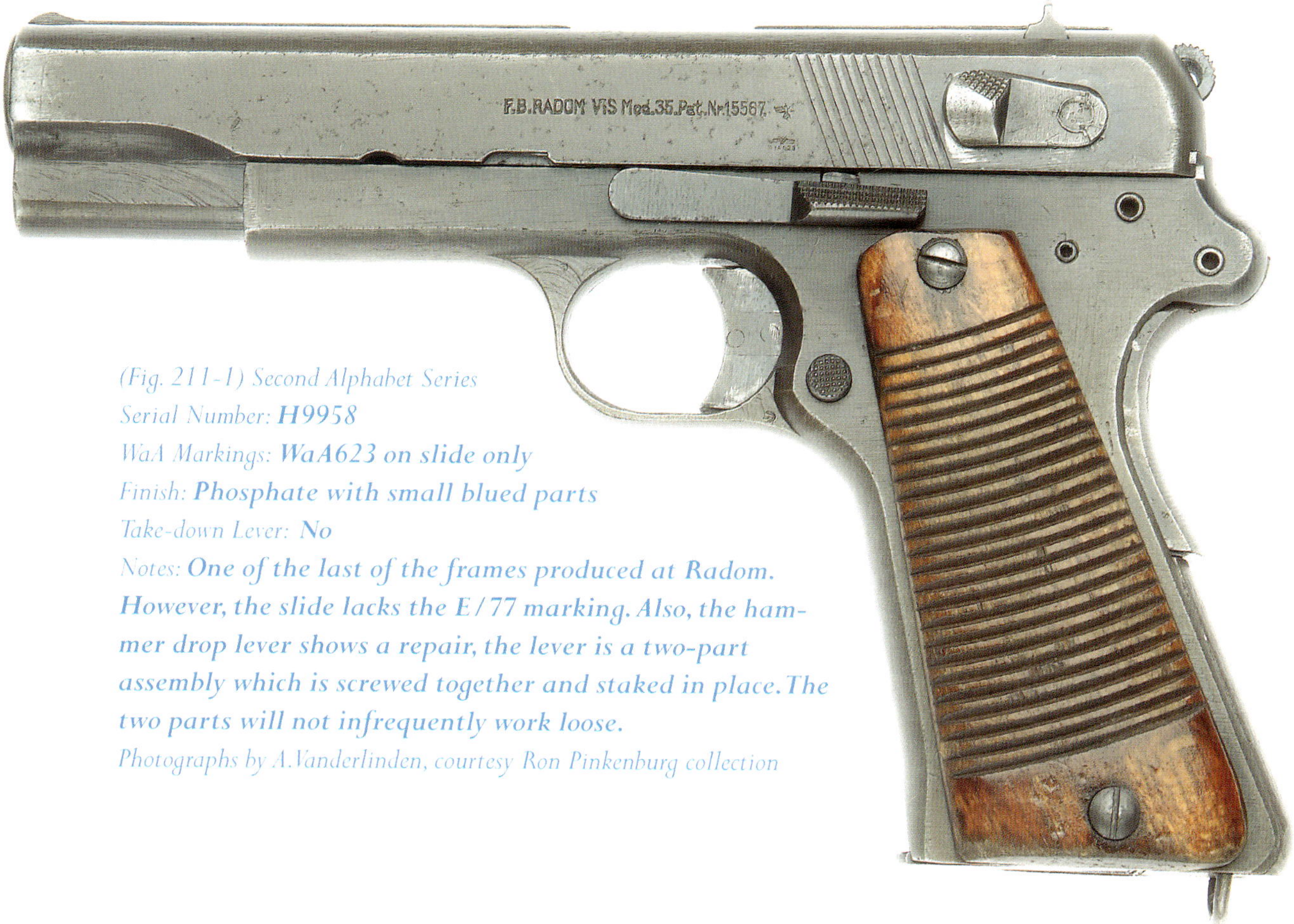

(Fig. 211-1) Second Alphabet Series
Serial Number: ***H9958***
WaA Markings: ***WaA623 on slide only***
Finish: ***Phosphate with small blued parts***
Take-down Lever: ***No***
Notes: ***One of the last of the frames produced at Radom. However, the slide lacks the E/77 marking. Also, the hammer drop lever shows a repair, the lever is a two-part assembly which is screwed together and staked in place. The two parts will not infrequently work loose.***
Photographs by A.Vanderlinden, courtesy Ron Pinkenburg collection

Occupation Production: Second alphabet prefix VIS N° J0660

(Fig. 212-1) Second Alphabet Series
Serial Number: ***J0660***
WaA Markings: ***WaA623***
Finish: ***Blue***
Take-down Lever: ***No***
Notes: ***Pistols in the (second) "J" series are quite rare. Apparently they were produced from the last batch of numbered parts manufactured at the Radom plant, and evacuated by train to Steyr when Soviet forces were approaching Radom. The evacuation trains were subjected to Soviet bombardment, as was the Znaim factory, resulting in the loss of many of the "J" parts kits. This pistol has the Radom E/77 on its right frame, and the Steyr WaA623 on its left slide and frame (trigger guard). The pistol was marked after the war with the East German "sunburst" marking, indicating VOPO (East German Police) use. The barrel and slide have also been externally marked during the postwar period with the pistol's serial number.*** *Photographs by A. Vanderlinden, courtesy Ron Pinkenburg collection*

Occupation Production: Second alphabet prefix VIS N° K1028

(Fig. 213-1) Second Alphabet Series
Serial Number: ***K1028***
WaA Markings: ***WaA623, E/623 on triggerguard***
Finish: ***Blue***
Take-down Lever: ***No***
Notes: ***Steyr/Znaim production. The pistol lacks the Radom E/77 marking. The lack of the E/77 marking indicates that the slide and frame were manufactured at Steyr/Znaim as part of the 1945 production. It is fitted with typical late war Steyr grooved wooden grips.***
Photographs by A. Vanderlinden, courtesy Waclaw Ustupski collection

Occupation Production: Second alphabet prefix VIS N° K1759

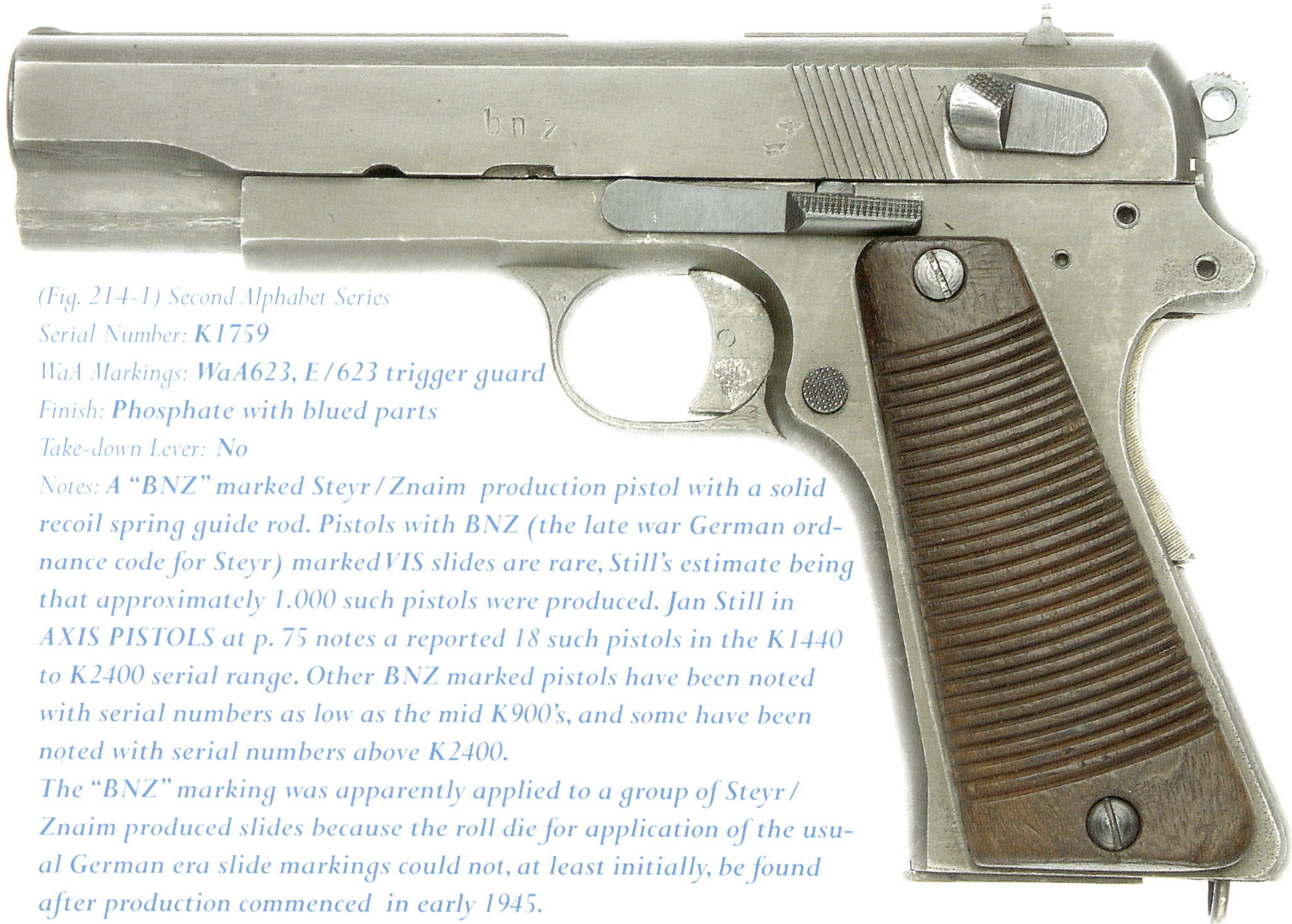

(Fig. 214-1) Second Alphabet Series
Serial Number: ***K1759***
WaA Markings: ***WaA623, E/623 trigger guard***
Finish: ***Phosphate with blued parts***
Take-down Lever: ***No***
Notes: ***A "BNZ" marked Steyr/Znaim production pistol with a solid recoil spring guide rod. Pistols with BNZ (the late war German ordnance code for Steyr) marked VIS slides are rare, Still's estimate being that approximately 1.000 such pistols were produced. Jan Still in AXIS PISTOLS at p. 75 notes a reported 18 such pistols in the K1440 to K2400 serial range. Other BNZ marked pistols have been noted with serial numbers as low as the mid K900's, and some have been noted with serial numbers above K2400.***
The "BNZ" marking was apparently applied to a group of Steyr/Znaim produced slides because the roll die for application of the usual German era slide markings could not, at least initially, be found after production commenced in early 1945.

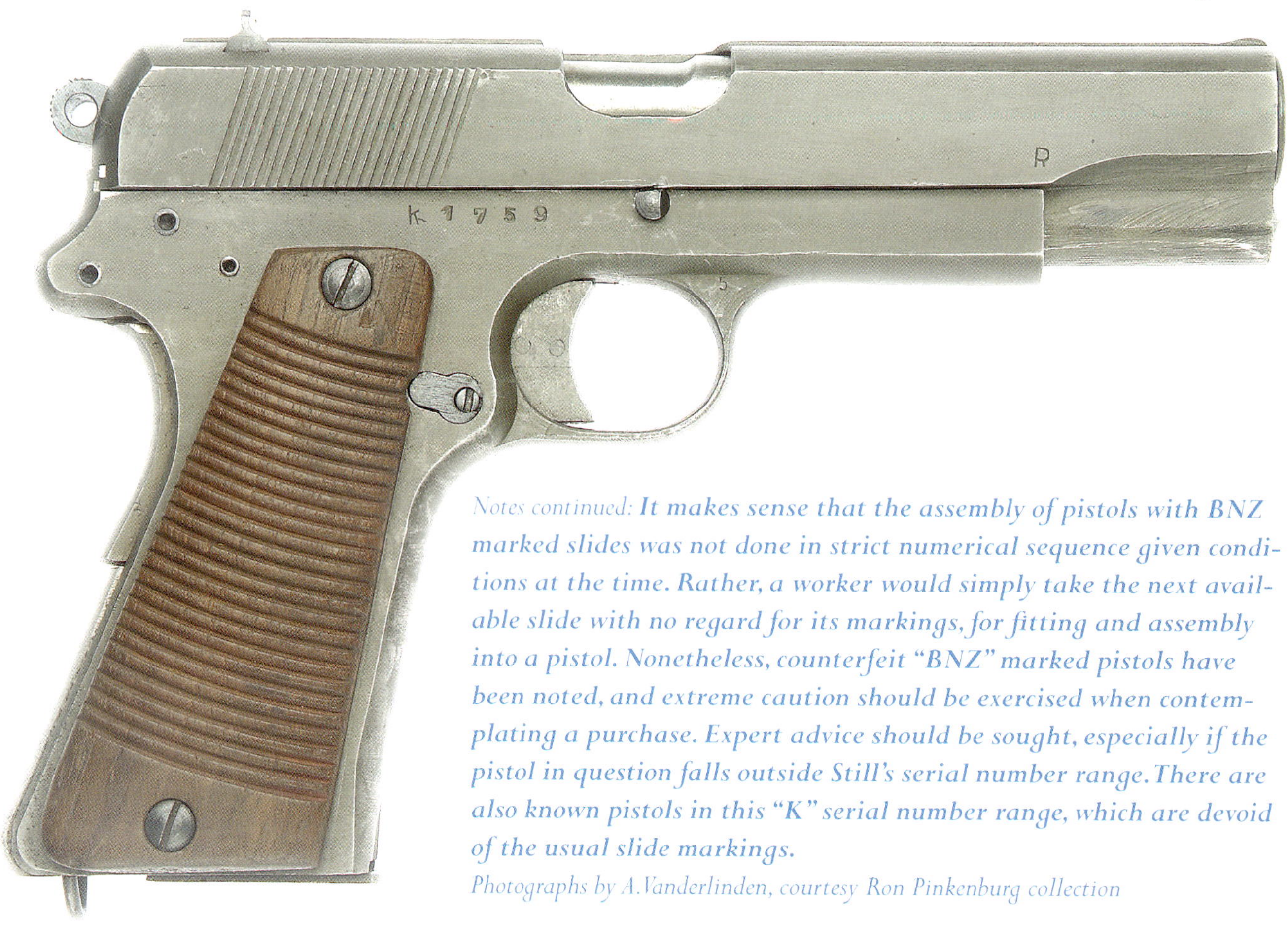

Notes continued: ***It makes sense that the assembly of pistols with BNZ marked slides was not done in strict numerical sequence given conditions at the time. Rather, a worker would simply take the next available slide with no regard for its markings, for fitting and assembly into a pistol. Nonetheless, counterfeit "BNZ" marked pistols have been noted, and extreme caution should be exercised when contemplating a purchase. Expert advice should be sought, especially if the pistol in question falls outside Still's serial number range. There are also known pistols in this "K" serial number range, which are devoid of the usual slide markings.***

Photographs by A.Vanderlinden, courtesy Ron Pinkenburg collection

Occupation Production: Second alphabet prefix VIS N° K1759 (continued)

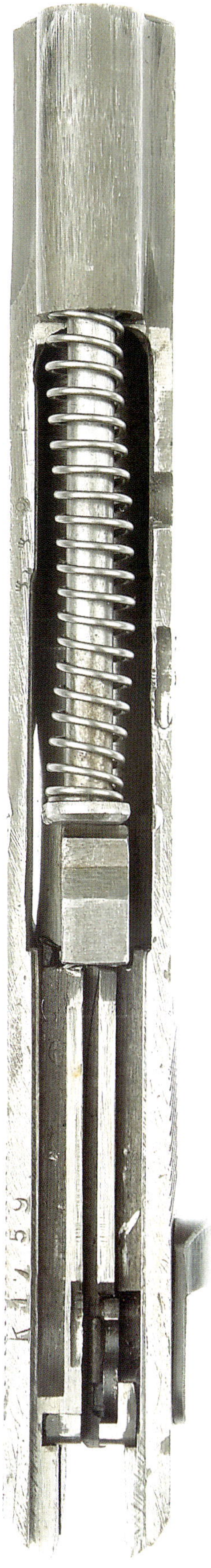

(Fig. 215-1) Notes: *"Illustrations of the BNZ pistol's slide inventory number, the hand stamping characteristic of the "BNZ" slide marking, and of the riveted pins characteristic of late war magazine floorplate attachment."*

(Fig. 215-2) Notes: *The underside of the BNZ pistol's slide. These late war slides were serial numbered on the bottom of the right slide rail. The author has seen a BNZ marked slide which had had its original slide number shaved off and replaced to match an "almost BNZ" serialized barrel and frame, producing a counterfeit.*

Occupation Production: Second alphabet prefix VIS Nº K2024

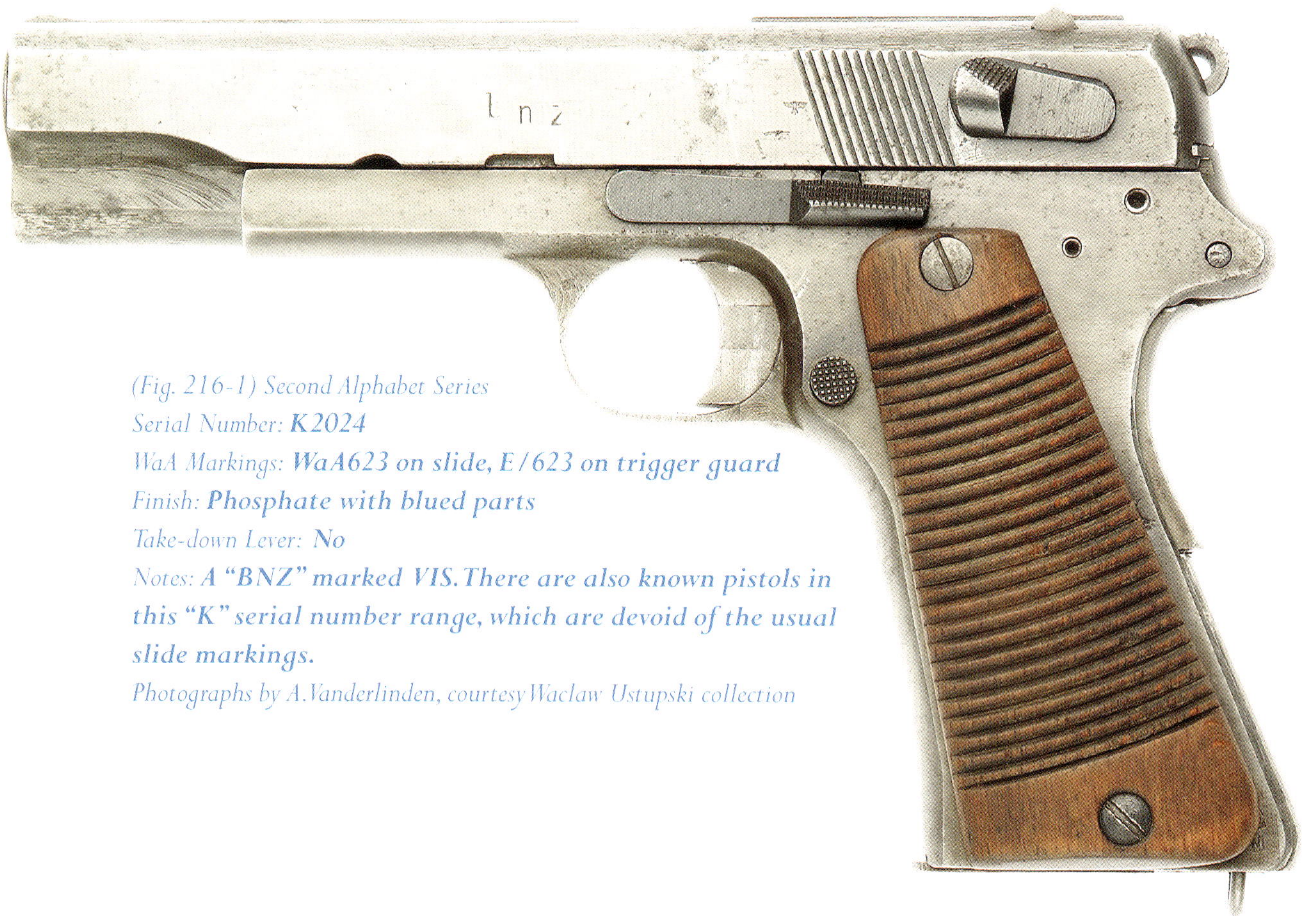

(Fig. 216-1) Second Alphabet Series
Serial Number: ***K2024***
WaA Markings: ***WaA623 on slide, E/623 on trigger guard***
Finish: ***Phosphate with blued parts***
Take-down Lever: ***No***
Notes: ***A "BNZ" marked VIS. There are also known pistols in this "K" serial number range, which are devoid of the usual slide markings.***
Photographs by A. Vanderlinden, courtesy Waclaw Ustupski collection

Occupation Production: Second alphabet prefix VIS N° K2177

(Fig. 217-1) Second Alphabet Series

Serial Number: ***K2177***

WaA Markings: ***E/623 on slide and triggerguard***

Finish: ***Phosphate with blued parts***

Take-down Lever: ***No***

Notes: ***A BNZ marked pistol with typical phosphate finish and blue parts. Note that the slide serrations were not milled, yet the pistol was proofed and accepted. Pistols in this serial range have been reported with missing or partial slide serrations. It is unknown why the slide serrations were not milled or why the standard slide legend was not used. These features have left a lot to speculation; from broken machinery to relocated production due to Allied bombing.***

Photographs by A.Vanderlinden, courtesy David Rachwal

Occupation Production: Second alphabet prefix VIS N° K4030

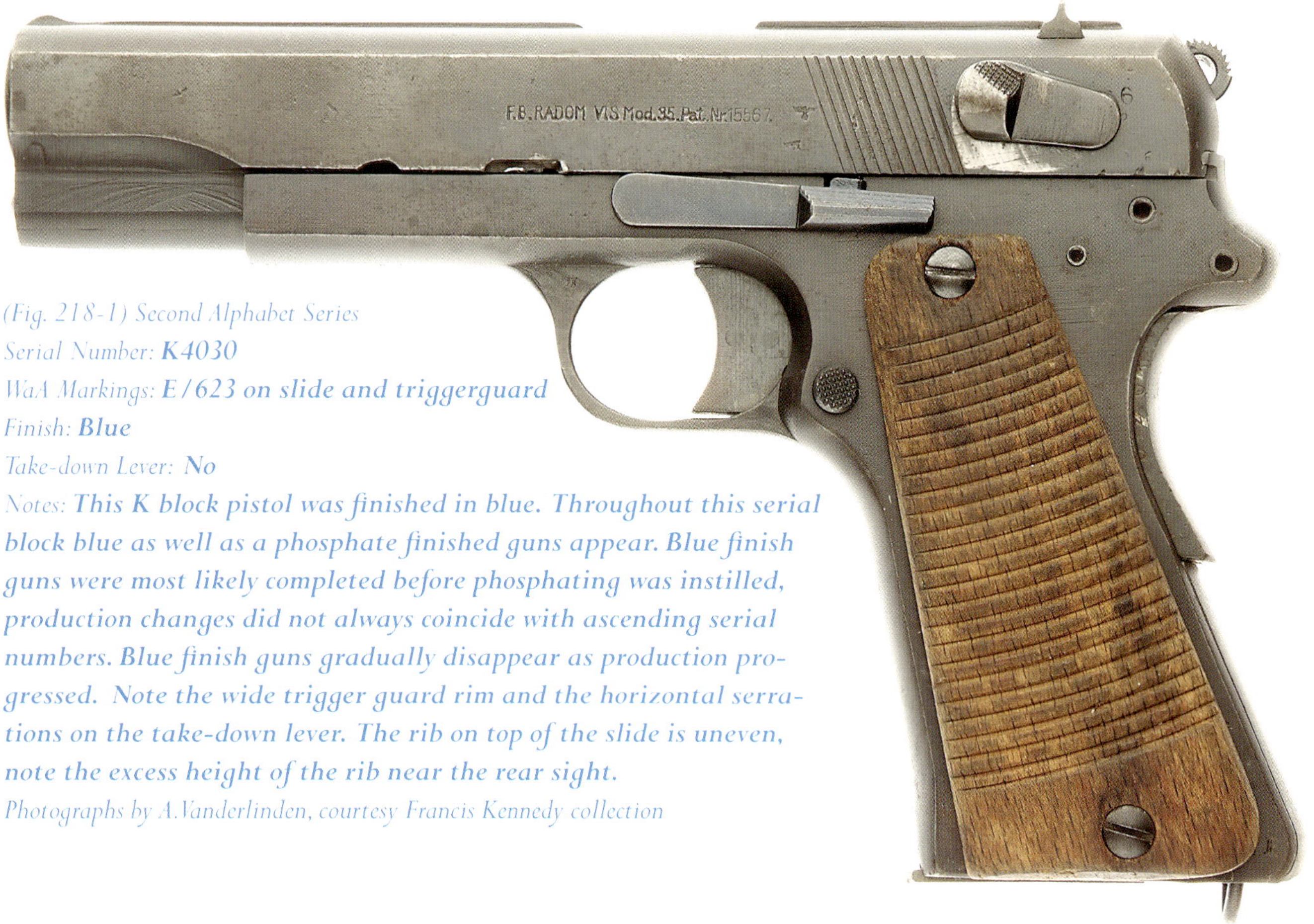

(Fig. 218-1) Second Alphabet Series
Serial Number: ***K4030***
WaA Markings: ***E/623 on slide and triggerguard***
Finish: ***Blue***
Take-down Lever: ***No***
Notes: ***This K block pistol was finished in blue. Throughout this serial block blue as well as a phosphate finished guns appear. Blue finish guns were most likely completed before phosphating was instilled, production changes did not always coincide with ascending serial numbers. Blue finish guns gradually disappear as production progressed. Note the wide trigger guard rim and the horizontal serrations on the take-down lever. The rib on top of the slide is uneven, note the excess height of the rib near the rear sight.***
Photographs by A.Vanderlinden, courtesy Francis Kennedy collection

Occupation Production: Second alphabet prefix VIS N° K5219

(Fig. 219-1) Second Alphabet Series

Serial Number: ***K5219***

WaA Markings: ***E/623 on slide and triggerguard***

Finish: ***Phosphate with blued parts***

Take-down Lever: ***No***

Notes: ***A late Steyr/Znaim produced pistol with the usual wartime slide marking. Note, however, the exceptionally thick and unfinished trigger guard. Note the comparison with pistol N° K7287. Note especially the lack of finishing on pistol N° K5219. The matching serial number on the grip safety is visible on the left side. A crude but functional pistol.***

Photographs by A.Vanderlinden, courtesy Waclaw Ustupski collection

Occupation Production: Second alphabet prefix VIS N° K5548

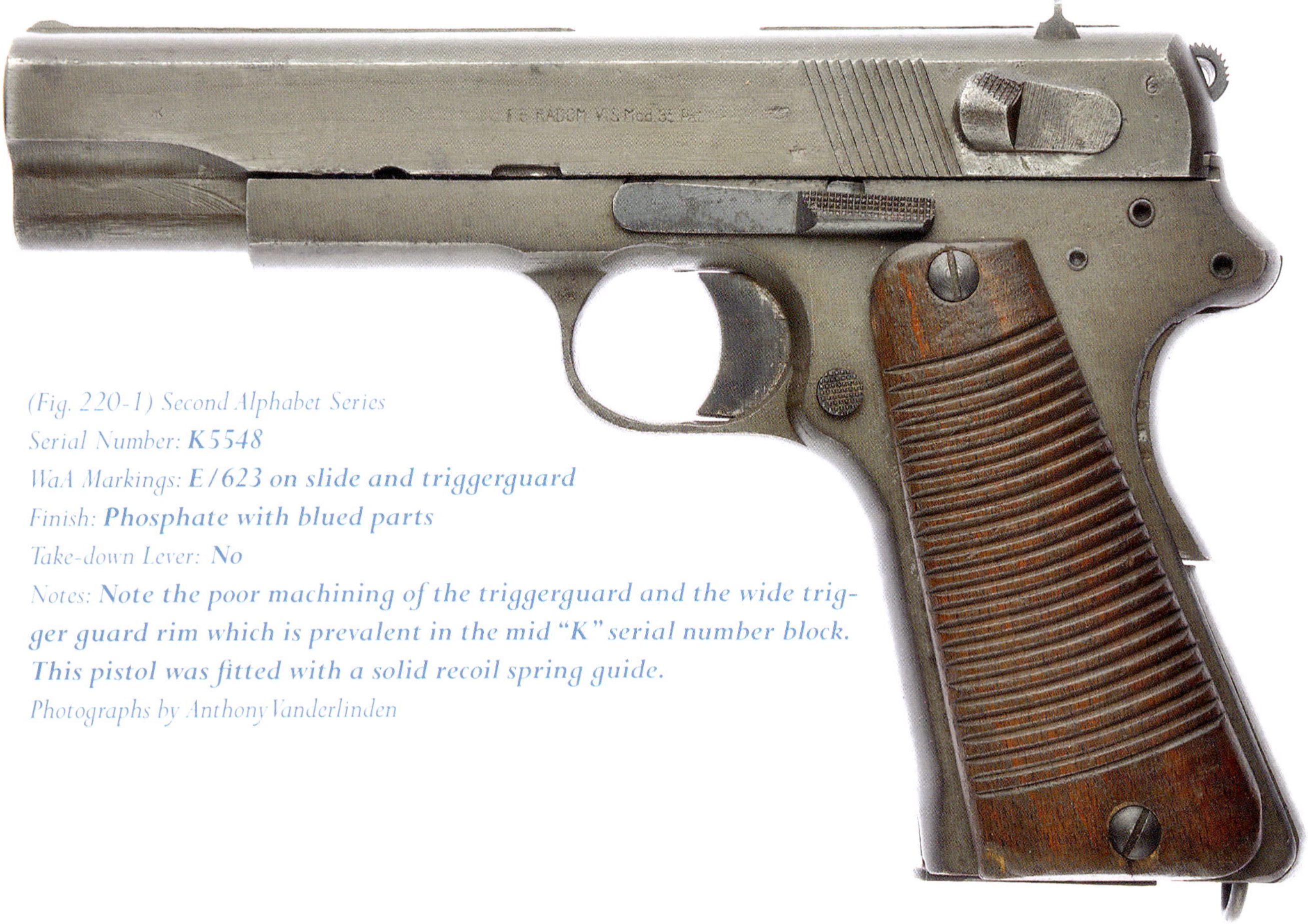

(Fig. 220-1) Second Alphabet Series
Serial Number: **K5548**
WaA Markings: **E/623 on slide and triggerguard**
Finish: **Phosphate with blued parts**
Take-down Lever: **No**
Notes: **Note the poor machining of the triggerguard and the wide trigger guard rim which is prevalent in the mid "K" serial number block. This pistol was fitted with a solid recoil spring guide.**
Photographs by Anthony Vanderlinden

Occupation Production: Second alphabet prefix VIS N° K6598

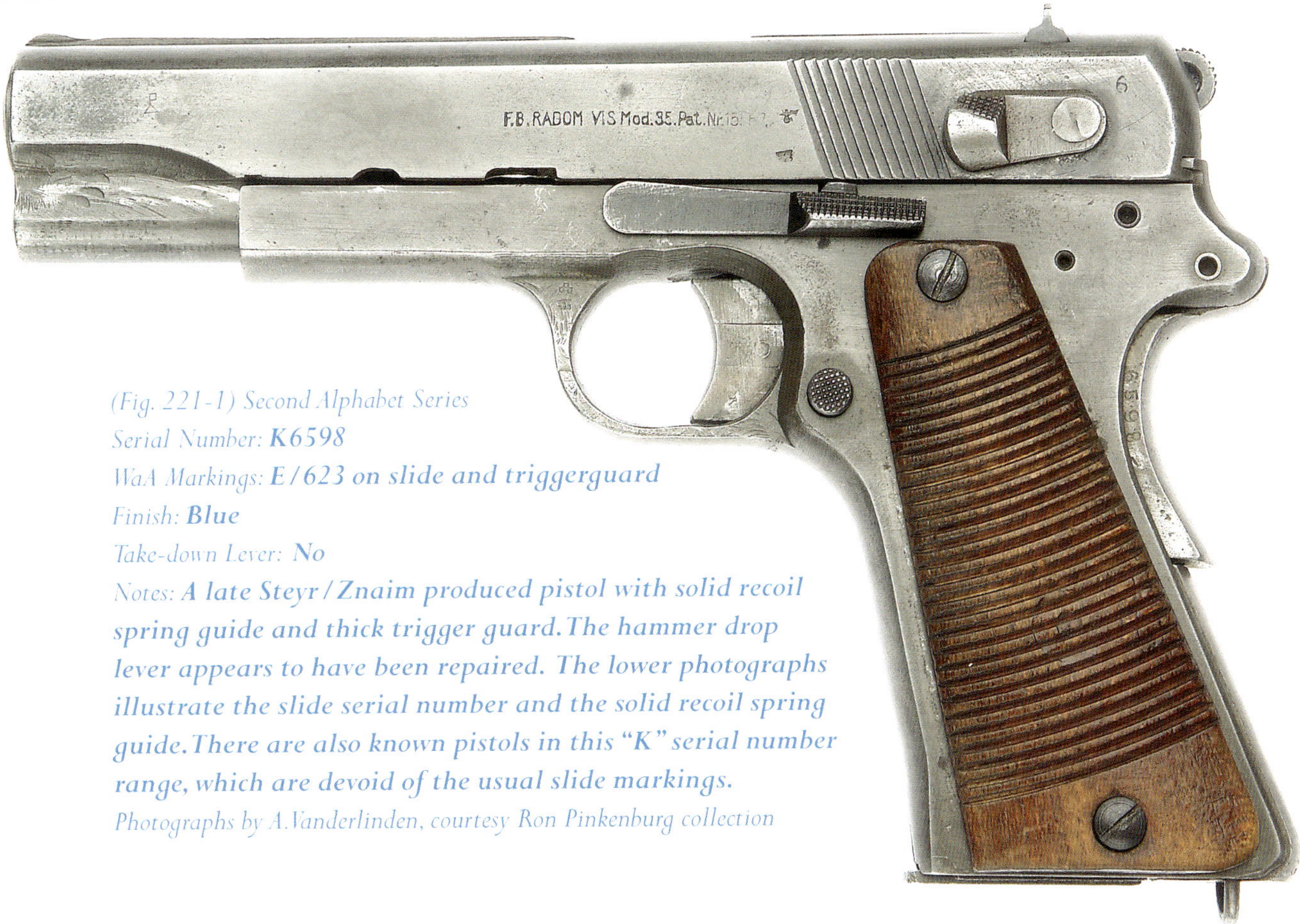

(Fig. 221-1) Second Alphabet Series
Serial Number: ***K6598***
WaA Markings: ***E/623 on slide and triggerguard***
Finish: ***Blue***
Take-down Lever: ***No***
Notes: ***A late Steyr/Znaim produced pistol with solid recoil spring guide and thick trigger guard. The hammer drop lever appears to have been repaired. The lower photographs illustrate the slide serial number and the solid recoil spring guide. There are also known pistols in this "K" serial number range, which are devoid of the usual slide markings.***
Photographs by A.Vanderlinden, courtesy Ron Pinkenburg collection

Occupation Production: Second alphabet prefix VIS N° K7287

(Fig. 222-1) Second Alphabet Series

Serial Number: **K7287**

WaA Markings: **E/623 on slide only**

Finish: **Phosphate with blued parts**

Take-down Lever: **No**

Notes: **A late Steyr/Znaim produced pistol. Note the visible serial number on the side of the grip safety, which is often seen on wartime production. Note also the steel defect in the slide which was ignored at the factory. The slide markings are very thin on this pistol. There are also known pistols in this "K" serial number range, which are devoid of the usual slide markings.**

Photographs by A. Vanderlinden, courtesy Waclaw Ustupski collection

Occupation Production: Second alphabet prefix VIS N° K8228

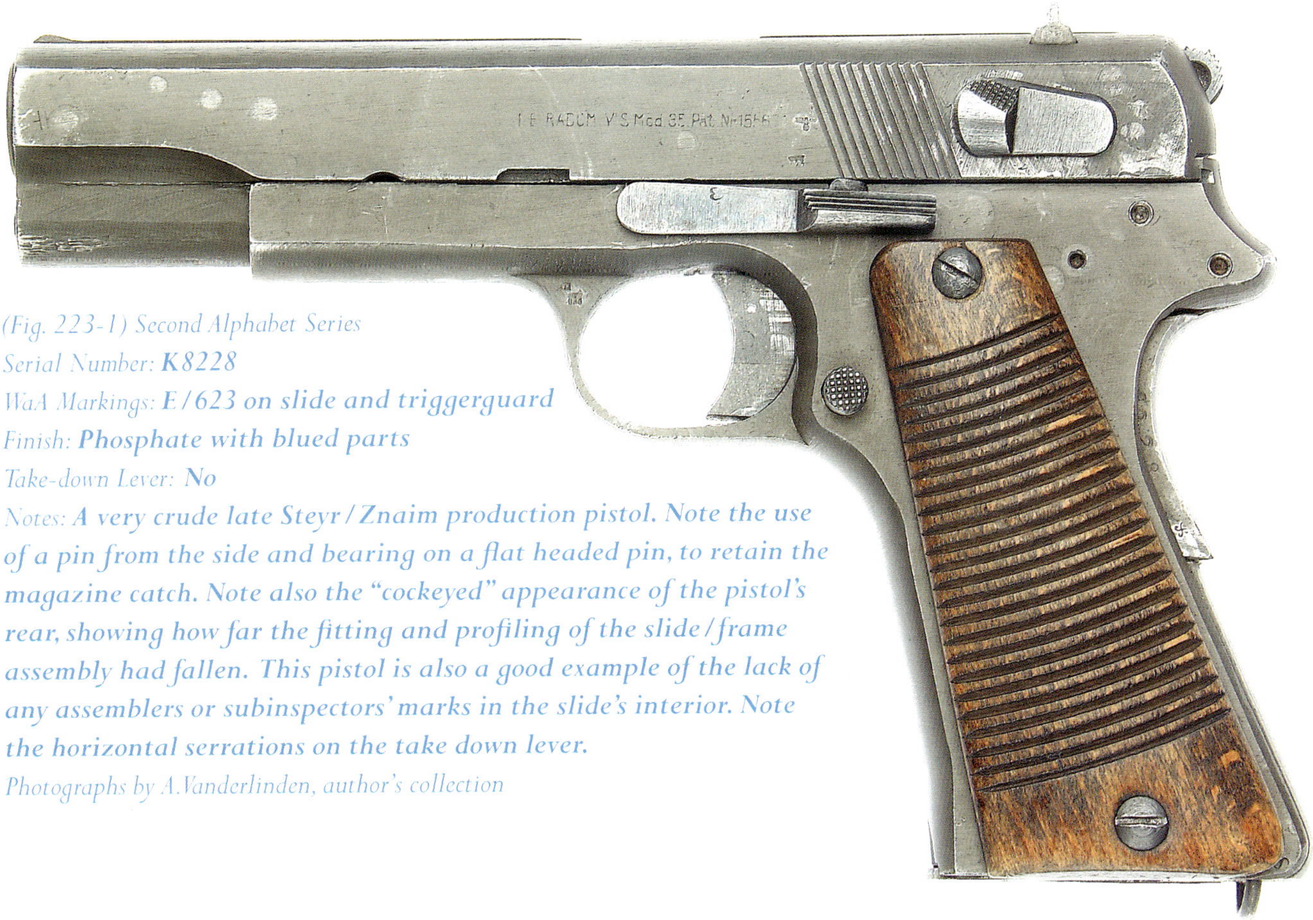

(Fig. 223-1) Second Alphabet Series
Serial Number: ***K8228***
WaA Markings: ***E/623 on slide and triggerguard***
Finish: ***Phosphate with blued parts***
Take-down Lever: ***No***
Notes: ***A very crude late Steyr/Znaim production pistol. Note the use of a pin from the side and bearing on a flat headed pin, to retain the magazine catch. Note also the "cockeyed" appearance of the pistol's rear, showing how far the fitting and profiling of the slide/frame assembly had fallen. This pistol is also a good example of the lack of any assemblers or subinspectors' marks in the slide's interior. Note the horizontal serrations on the take down lever.***
Photographs by A.Vanderlinden, author's collection

(Fig. 224-1) Notes: **Illustration of the ersatz method of retaining the magazine catch and spring.**

(Fig. 224-2) Notes: **This shows the "cock-eyed" (and non interchangeable) fitting of the pistol's grip safety and mainspring housing in its frame. Note the horizontal serrations on the take down lever.**

(Fig. 224-3) Notes: **The slide bottom has a serial number, but no other markings.**

Occupation Production: Second alphabet prefix VIS N° K8453

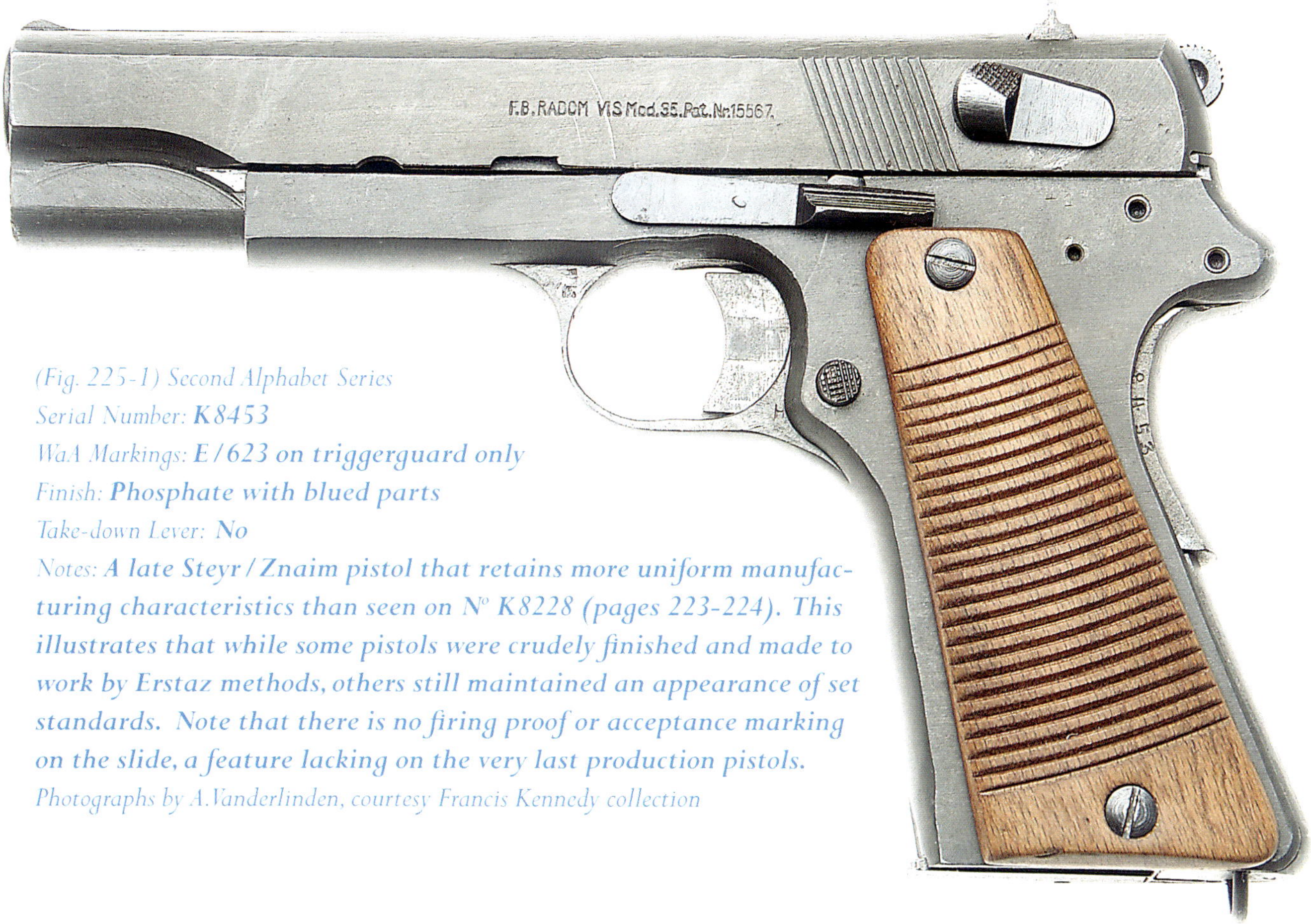

(Fig. 225-1) Second Alphabet Series
Serial Number: ***K8453***
WaA Markings: ***E/623 on triggerguard only***
Finish: ***Phosphate with blued parts***
Take-down Lever: ***No***
Notes: ***A late Steyr/Znaim pistol that retains more uniform manufacturing characteristics than seen on N° K8228 (pages 223-224). This illustrates that while some pistols were crudely finished and made to work by Erstaz methods, others still maintained an appearance of set standards. Note that there is no firing proof or acceptance marking on the slide, a feature lacking on the very last production pistols.***
Photographs by A.Vanderlinden, courtesy Francis Kennedy collection

Occupation Production: Second alphabet prefix VIS N° K9132

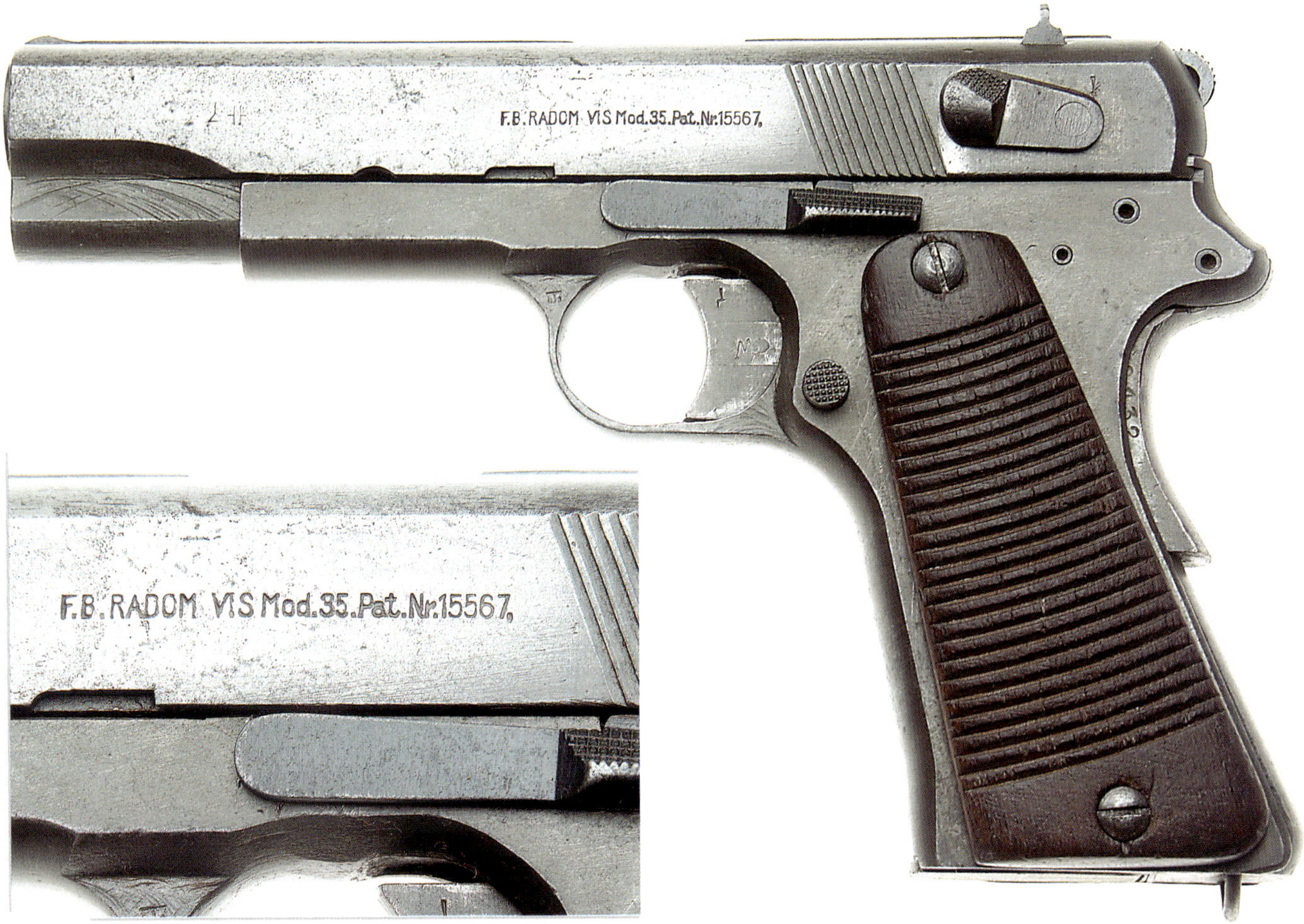

(Fig. 226-1) Second Alphabet Series

Serial Number: ***K9132***

WaA Markings: ***E/623 on triggerguard only***

Finish: ***Phosphate with blued parts***

Take-down Lever: ***No***

Notes: ***One of the last pistols produced. The highest number reported by a contributor was N° 9150. Note that there is no firing proof or acceptance marking on the slide, a feature lacking on the very last production pistols.***

Photographs by A. Vanderlinden, courtesy Francis Kennedy collection

Photography Gallery IV

VIS Pistols of the Polish Underground

(Fig. 228-1) Polish Underground VIS
Serial Number: ***None***
German Markings: ***None***
Slotted for Shoulder-Stock: ***No***
Take-down Lever: ***No***
Notes: ***An underground pistol made of late war parts, with riveted pins, but with small grip screws and bushings. The slide and frame both have the typical late war inventory numbers. The pistol has a prewar Polish barrel with serial number 1992. The front left of the frame is marked "AJ". This is an unknown marking.***
Photographs by A. Vanderlinden, courtesy Ed Tytor collection

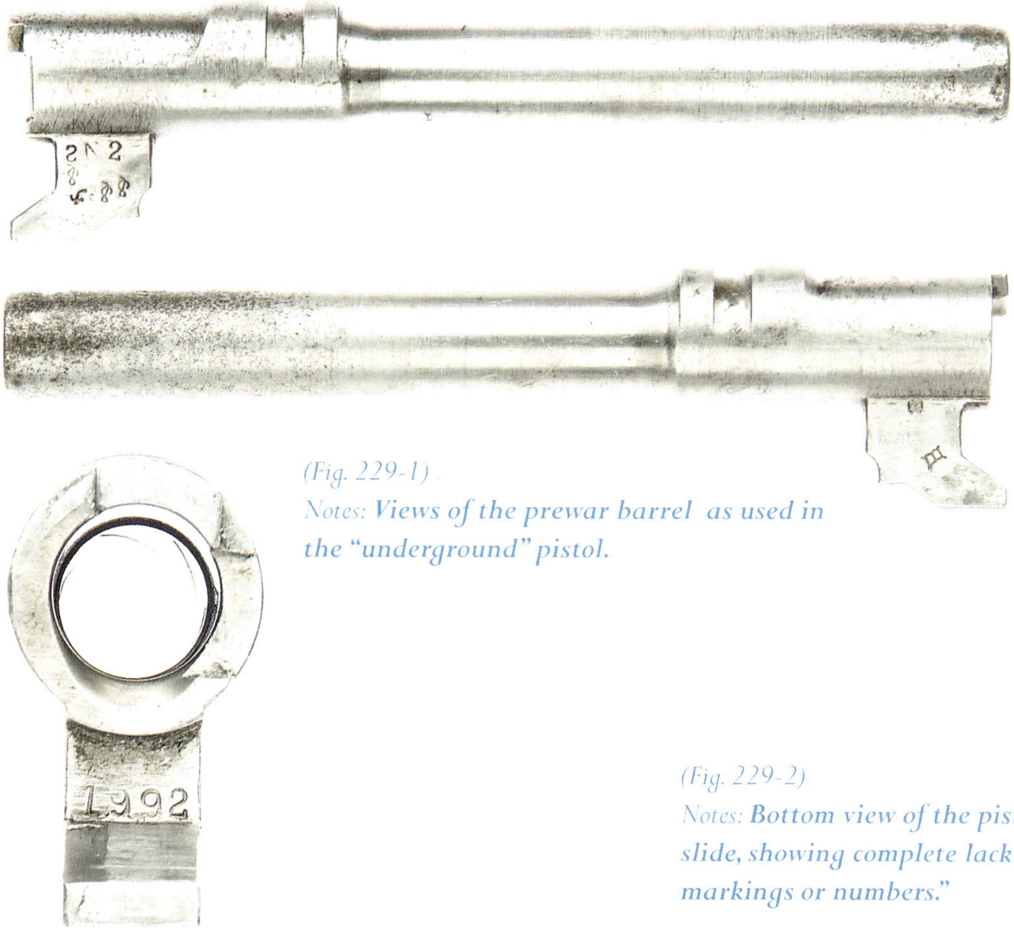

(Fig. 229-1)
Notes: ***Views of the prewar barrel as used in the "underground" pistol.***

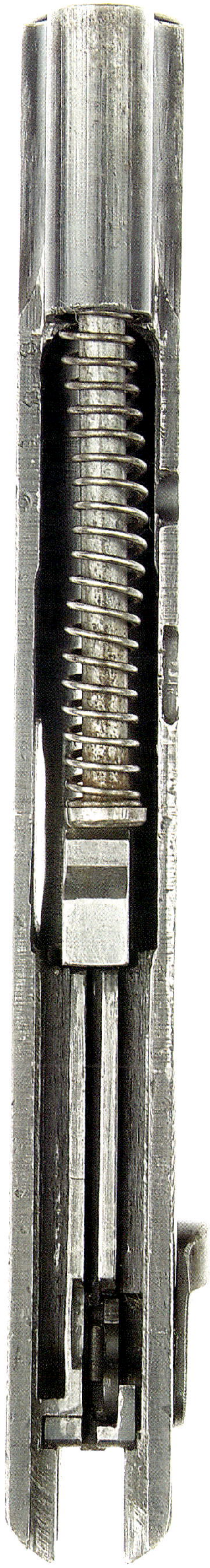

(Fig. 229-2)
Notes: ***Bottom view of the pistol's slide, showing complete lack of markings or numbers."***

(Fig. 229-3)
Notes: ***Slide inventory number.***

(Fig. 230-1) Polish Underground VIS
Serial Number: ***158***
German Markings: ***None***
Slotted for Shoulder-Stock: ***No***
Take-down Lever: ***No***
Notes: ***This has all the appearances of a very late pistol, with a fabricated trigger and riveted frame pins. The pistol shows no waffenamt markings, and has an odd bulging left grip. It is the author's belief that the serial number on this pistol may be a non-factory number applied to a gun made of stolen parts and intended for use by the Polish underground. An incident is reported (MACEJEWSKI, p. 170) of a Polish underground pistol bearing serial number 1078, which was discovered by the Gestapo, resulting in the arrest of 80 Radom employees for conspiracy to produce underground weapons, ending in the execution of 50 employees.***
Photographs by A. Vanderlinden, courtesy Ron Pinkenburg collection

Polish Underground Pistols: VIS N° -

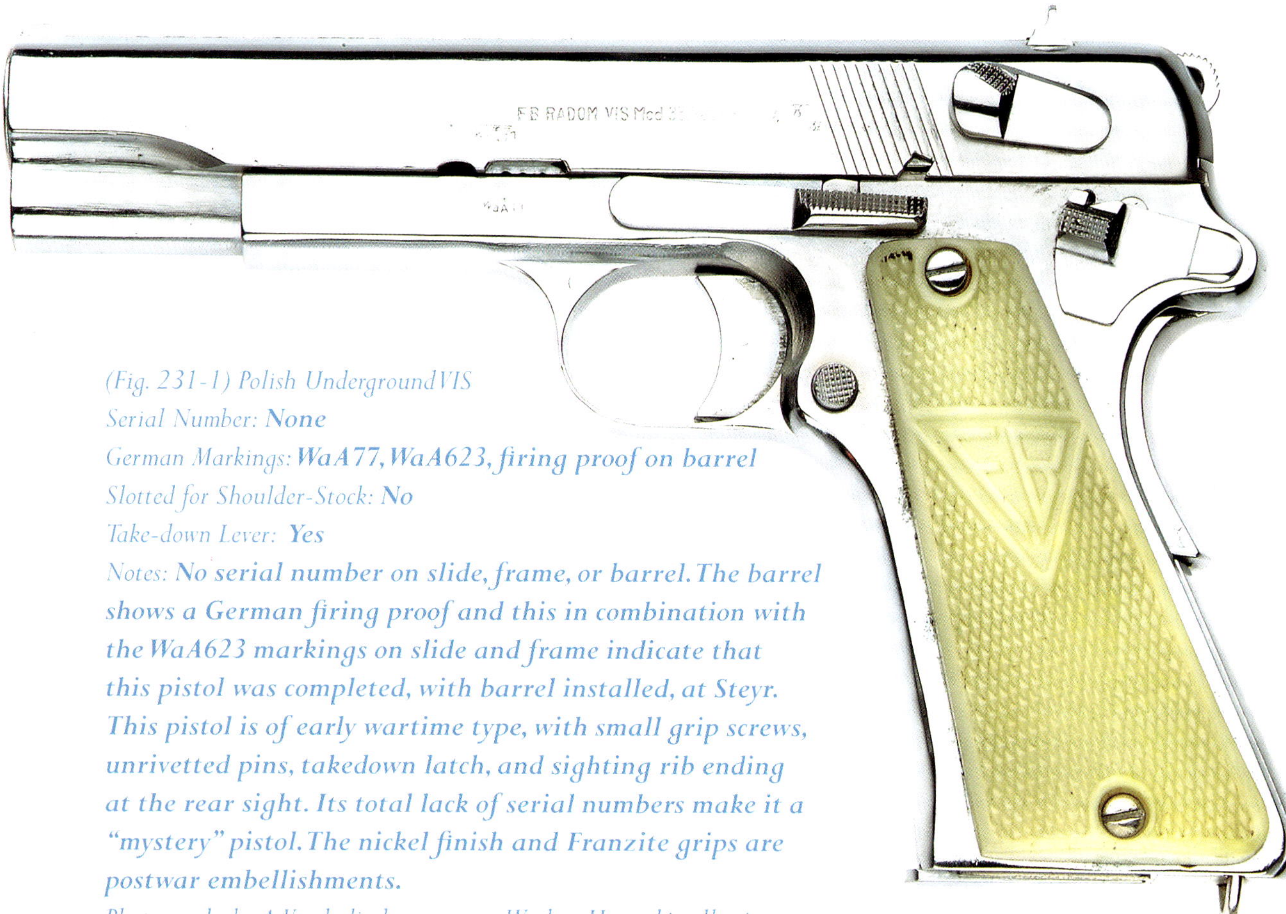

(Fig. 231-1) Polish Underground VIS
Serial Number: ***None***
German Markings: ***WaA77, WaA623, firing proof on barrel***
Slotted for Shoulder-Stock: ***No***
Take-down Lever: ***Yes***
Notes: ***No serial number on slide, frame, or barrel. The barrel shows a German firing proof and this in combination with the WaA623 markings on slide and frame indicate that this pistol was completed, with barrel installed, at Steyr. This pistol is of early wartime type, with small grip screws, unrivetted pins, takedown latch, and sighting rib ending at the rear sight. Its total lack of serial numbers make it a "mystery" pistol. The nickel finish and Franzite grips are postwar embellishments.***
Photographs by A. Vanderlinden, courtesy Waclaw Ustupski collection

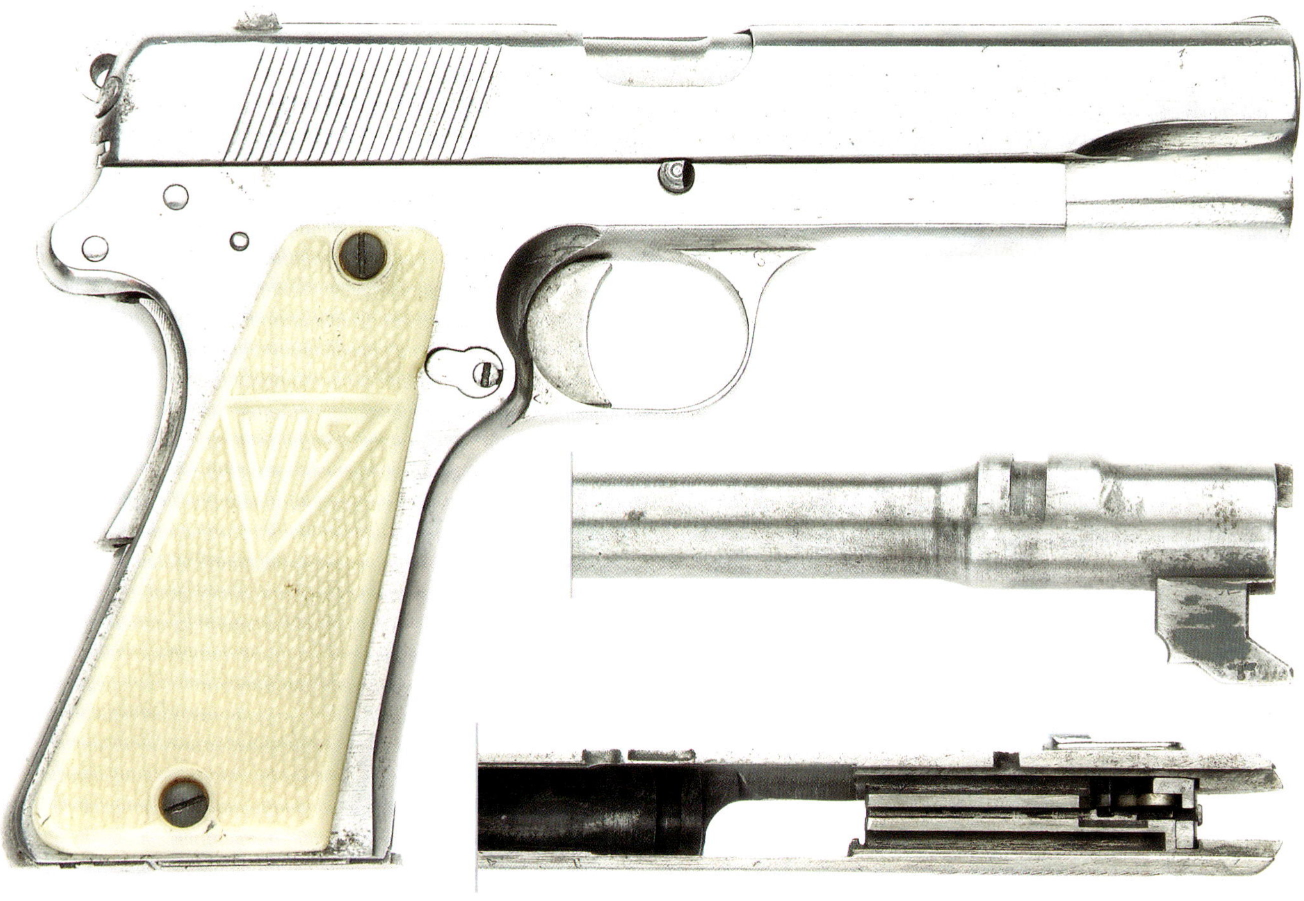

Photography Gallery V

Non-VIS Handgun Production at Radom

Non-Vis Production Handguns: 1931 Ng30 (Nagant) Revolver

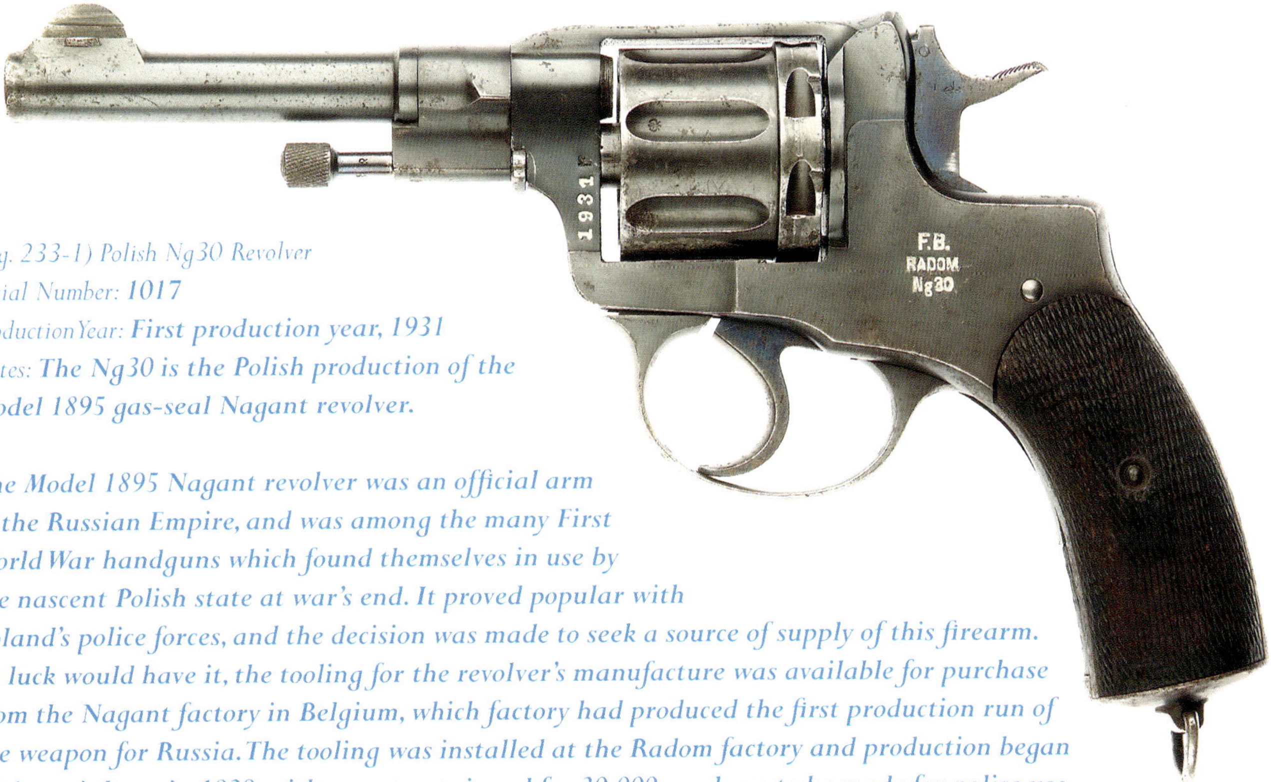

(Fig. 233-1) Polish Ng30 Revolver
Serial Number: ***1017***
Production Year: ***First production year, 1931***
Notes: ***The Ng30 is the Polish production of the Model 1895 gas-seal Nagant revolver.***

The Model 1895 Nagant revolver was an official arm of the Russian Empire, and was among the many First World War handguns which found themselves in use by the nascent Polish state at war's end. It proved popular with Poland's police forces, and the decision was made to seek a source of supply of this firearm. As luck would have it, the tooling for the revolver's manufacture was available for purchase from the Nagant factory in Belgium, which factory had produced the first production run of the weapon for Russia. The tooling was installed at the Radom factory and production began with a trial run in 1929, with a contract signed for 30,000 revolvers to be made for police use.

There is some confusion as to the actual number of Ng30 revolvers produced in the production run lasting from 1931 to 1937. This may be traced to a Polish source's statement that only 7166 revolvers were produced from 1931 to 1935. Actually, this figure does not appear to count 1935 production, as revolvers in the mid 15000 serial number range and dated 1935 are known to exist, and a reported 1937 pistol is only in the 17000 serial number range, it seems that the original contract for 30000 was not met. The best guess is a total production of some 20000 revolvers.

Notes continued: ***The Ng30 Nagant is, like its Russian predecessor, a 7-shot revolver chambered for a unique 7.62mm cartridge and designed with a "gas seal" mechanism whereby the cylinder is shifted forward prior to firing enabling the front of the cartridge case to enter into the rear of the barrel and thus avoiding any escape of firing gases from the barrel-cylinder junction.***
Photographs by A. Vanderlinden, courtesy Waclaw Ustupski collection

(Fig. 234-1) Polish Ng30 Revolver
Serial Number: ***5499***
Production Year: ***1932***
Notes: ***See notes on page 233.***
Photographs by A. Vanderlinden, courtesy Francis Kennedy collection

Non-Vis Production Handguns: 1934 Ng30 (Nagant) Revolver

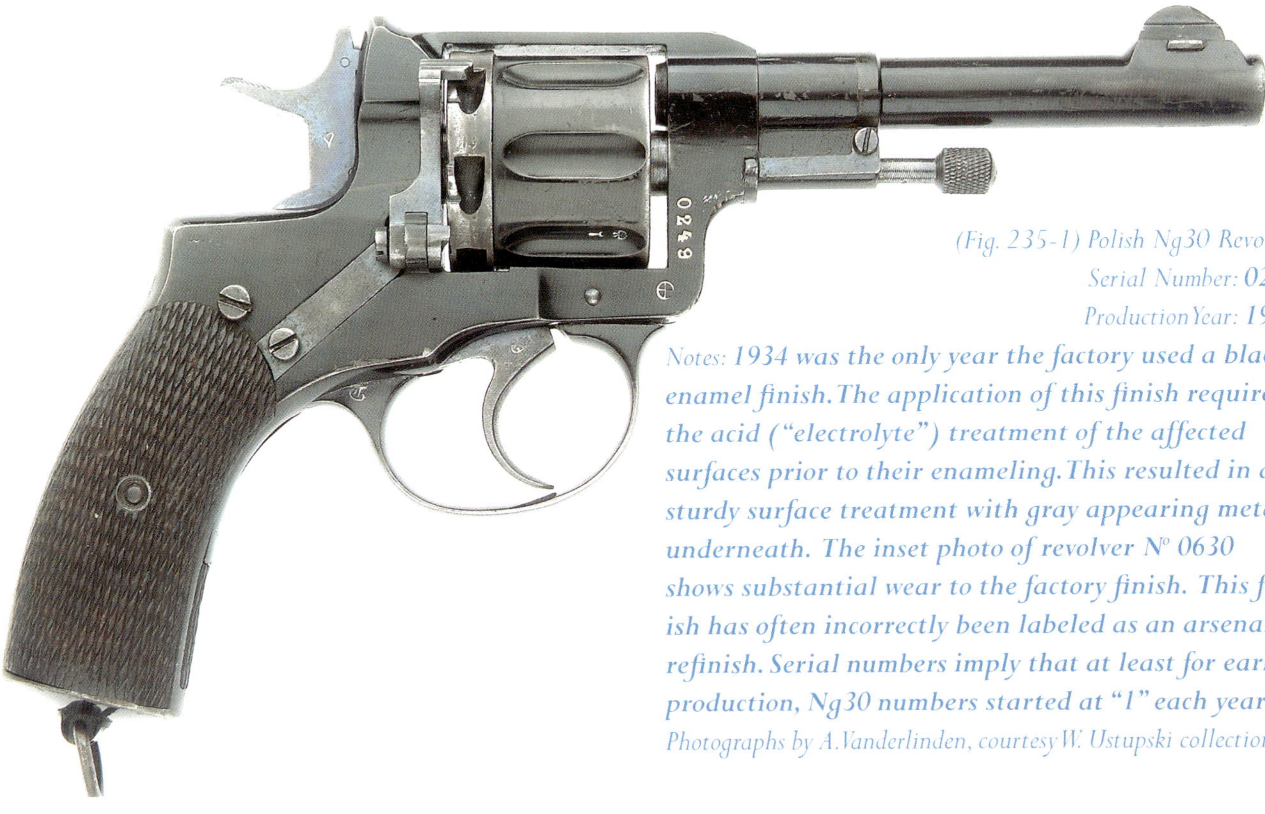

(Fig. 235-1) Polish Ng30 Revolver
Serial Number: 0249
Production Year: 1934

Notes: 1934 was the only year the factory used a black enamel finish. The application of this finish required the acid ("electrolyte") treatment of the affected surfaces prior to their enameling. This resulted in a sturdy surface treatment with gray appearing metal underneath. The inset photo of revolver N° 0630 shows substantial wear to the factory finish. This finish has often incorrectly been labeled as an arsenal refinish. Serial numbers imply that at least for early production, Ng30 numbers started at "1" each year.

Photographs by A. Vanderlinden, courtesy W. Ustupski collection

(Fig. 236-1) Polish Ng30 Revolver
Serial Number: ***8425***
Production Year: ***1935***
Notes: **See notes on page 233.**
Photographs by A.Vanderlinden, courtesy Ron Pinkenburg collection

Non-Vis Production Handguns: .22 caliber Sportowy

(Fig. 237-1) Polish Sportowy
Serial Number: ***S1918***
Production Year: ***1955***
Notes: ***After the conclusion of World War II, with Poland in the Soviet orbit, pistol production at Radom was restarted with a Polish produced version of the TT33 Tokarev automatic pistol. The Polish Tokarev trainer, the .22 caliber Sportowy, followed. Note especially the cartridge adapters, which hold the .22 round and are designed to feed through the standard military Tokarev magazine. Compare these to the illustrated barrel insert and cartridge holder for the projected .22 VIS conversion of 1939.***
Photographs by A. Vanderlinden, courtesy Waclaw Ustupski collection

(Fig. 238-1) Polish P64 (Makarov)
Serial Number: ***ES07790***
Production Year: ***1973***
Notes: ***Designs for pistols chambering the 9mm Makarov cartridge, like the P64, followed production of the TT33 Tokarev. Quite a few of these blowback operated pistols have been imported into the United States. The number "11" in an oval on the right side of the slide is the former Warsaw Pact's country code for Poland.***
Photographs by A. Vanderlinden, courtesy Waclaw Ustupski collection

Non-Vis Production Handguns: P83 (Makarov)

(Fig. 239-1) Polish P83 (Makarov)
Serial Number: ***7U09132***
Production Year: ***1991***
Notes: ***The P83 was a modernized version of the P64 using stamped steel parts. The number "11" in an oval on the right side of the slide is the former Warsaw Pact's country code for Poland.***
Photographs by A. Vanderlinden, courtesy Waclaw Ustupski collection

(Fig. 240-1) Polish MAG95
Serial Number: ***W00602***
Production Year: ***1997***
Notes: ***Polish MAG95, a Post-Communist pistol in caliber 9mm Parabellum. Like many other locked breech pistols of its generation, it has a captive recoil spring and is locked and unlocked by means of camming surfaces on the underbarrel lug.***
Photographs by A.Vanderlinden, courtesy Waclaw Ustupski collection

Photography Gallery VI

Rifle Production at Radom

Radom Rifle Production: Polish Karabinek (KBK) wz.1898

(Fig. 242-1) Polish Karabinek (KBK) wz.1898
Serial Number: ***7317***
Production Year: ***1927***
Caliber: ***7.92x57mm***
Notes: ***This, the Polish version of the World War I German Model 98AZ short rifle, is similar to the German weapon save for its markings and the Polish stacking rod, which is made in two pieces with a 90 degree angle between its root and the attached rod. These short rifles were produced on ex-Danzig Royal Arsenal machinery, in Warsaw from 1925 and 1931, with a total produced of 189,600. These rifles are marked with the legend of the Warsaw State Rifle Factory, PFK Warszawa, and the year produced. Production of this arm at Radom took place from April of 1927 to 1931, with a total Radom production of 158,100. This particular rifle is of 1927 Radom production and bears a receiver ring marking of "PWB Radom 1927". "PWB" stands for "Panstwowa Wytwornia Broni", meaning "State Weapon Factory". This abbreviation was replaced with the familiar "FB" (Weapon Factory) in 1929. The term "Karabinek" is a diminutive form of "Karabin", which latter refers to long rifles in Polish usage. "Karabinek" refers to carbines and short rifles. Polish short rifles were manufactured with both straight and bent bolt handles. Rifles with straight bolt handles were issued to the infantry, rifles with bent bolt handles, to the cavalry. All of the wooden buttstocks manufactured for these weapons were made with an inlet for the bent bolt handle. The stocks and handguards of these rifles will be found to have lateral strengthening dowels of a darker wood inlet behind the magazine box and, for the handguard, forward of and behind the rear sight. This is a characteristic of prewar Polish military rifle stocks.***
Photographs by A. Vanderlinden, courtesy Waclaw Ustupski collection

Radom Rifle Production: Polish Model 98

(Fig. 243-1) Polish Model 98
Serial Number: ***43681***
Production Year: ***1938***
Caliber: ***7.92x57mm***
Notes: ***The Polish version of the German Gewehr 98 long rifle was termed "Karabin 98" in its initial production run of 1922 and 1923. At that time some 22,000 of these rifles were produced at the Warsaw State Rifle Factory on ex-Danzig Royal Arsenal machinery acquired as a result of the Treaty of Versailles. The rear sights of at least the first of these rifles were the Lange sight used on World War I German Mauser long rifles. The Polish stocks were of Beechwood. The beechwood stocks of the wz. 98a will be found to have a lateral strengthening dowel of a darker wood inlet behind the magazine box. A second production run of Mauser long rifles, termed "Karabin Mausera wz. 1898a" took place at Radom from 1936 to 1939, with a volume of 44,500 produced. The latter rifles were made with tangent rear sights, as seen on the illustrated rifle.***
Photographs by A. Vanderlinden, courtesy Waclaw Ustupski collection

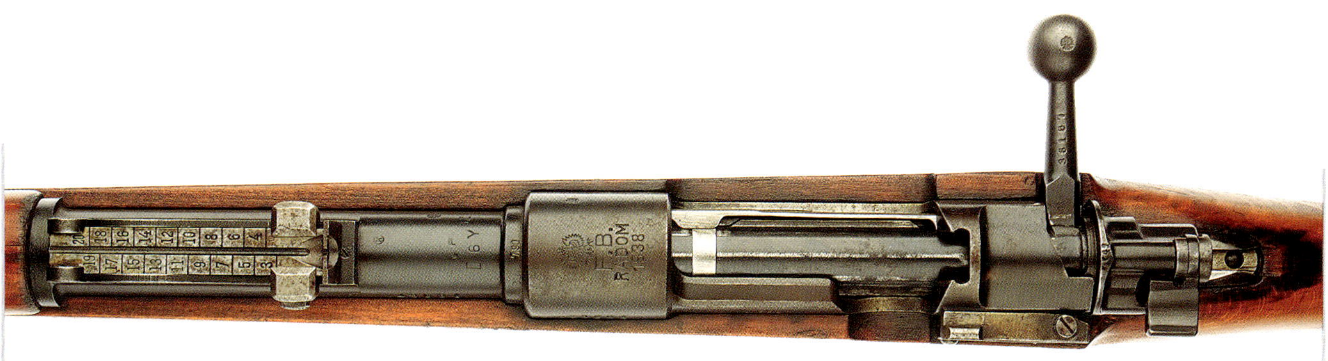

Radom Rifle Production: Karabinek wz 1929

(Fig. 244-1) Karabinek wz 1929

Caliber: ***7.92x57mm***

Notes: ***The Polish Model 1929 Short Rifle is one of a number of more or less contemporary European short rifle designs, built along similar principles. This rifle was produced at the Radom factory, with prototypes made in 1929 and full production taking place from 1930 to 1939. Total production was 264,300 rifles, which number includes an export version known as the wz. 29e. The latter lacked the usual receiver ring markings, having none. It was exported to Spain, which was then involved in a civil war, and to Afghanistan. Total exports are said to have been 7,900.***

Polish short rifles were manufactured with both straight and bent bolt handles. Rifles with straight bolt handles were issued to the infantry, rifles with bent bolt handles, to the cavalry. All of the wooden buttstocks manufactured for these weapons were made with an inlet for the bent bolt handle. The stocks and handguards of these rifles will be found to have lateral strengthening dowels of a darker wood inlet behind the magazine box and, for the handguard, forward of and behind the rear sight. This is a characteristic of prewar Polish military rifle stocks.

Photographs by A. Vanderlinden, courtesy Waclaw Ustupski collection

Radom Rifle Production: wz.31 .22 Caliber Trainer

(Fig. 245-1) wz.31 Training Rifle
Serial Number: ***3673***
Production Year: ***1932***
Caliber: ***.22***
Notes: ***This is a single-shot training rifle in caliber .22 Long Rifle, manufactured in the pre-World War II period by the Radom factory. Except for its distinctive bolt, the rifle is constructed to resemble the wz. 29 Mauser Short Rifle. Previous Polish .22 training rifles had been single-shot weapons with Mauser actions and closely resembling the KBK 1898 and the wz. 29. The stock and handguard of this rifle are, as is typical of prewar Polish military rifles, laterally dowelled with dark hardwood strengthening dowels.***
Photographs by A. Vanderlinden, courtesy Waclaw Ustupski collection

(Fig. 246-1) 1938 Maroszek Semi-Auto Rifle
Serial Number: ***1030***
Production Year: ***1938***
Caliber: ***7.92x57mm***
Notes: ***The M1938 Maroszek semiautomatic rifle was produced in prototype series only, the outbreak of war in 1939 preventing the start of planned series production. Of a total production of perhaps 150 rifles, only 5 are known to survive worldwide. This rifle is a 10-shot gas operated weapon in caliber 8x57 mm., with a non-detachable magazine loaded individually or by 5-round Mauser stripper clips from the top of the receiver. The rifle employs a tipping bolt system, the rear of the bolt tipping upward to lock in the top of the receiver. The bolt carrier and operating rod are a single unit with the gas piston running below the barrel and bolt. Externally the rifle reminds one of the Browning Automatic Rifle.***
Photographs by A. Vanderlinden, courtesy Waclaw Ustupski collection

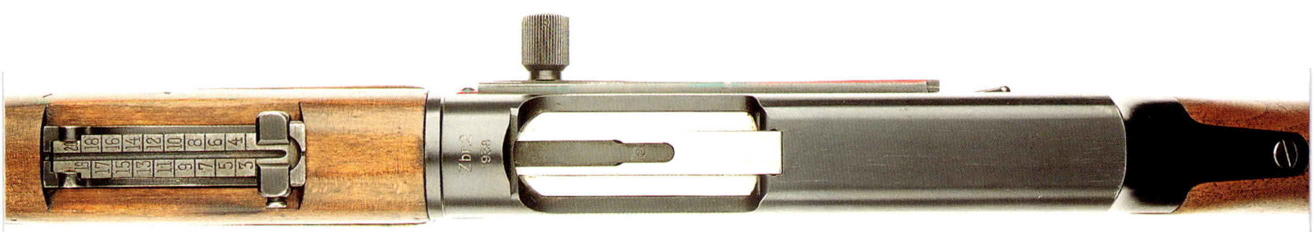

(Fig. 246-2) Notes: ***Top of rifle showing bolt with extractor and charging handle***
Photographs by A. Vanderlinden, courtesy Waclaw Ustupski collection

(Fig. 246-3) Notes: ***Front of rifle showing short forward part of stock and handguard, and gas tube under barrel***
Photographs by A. Vanderlinden, courtesy Waclaw Ustupski collection

(Fig. 247-1) Notes: ***Right and left side views of the Maroszek receiver. The upper photograph shows the charging handle, safety lever, and takedown lever (forward under the charging handle)***

Photographs by A. Vanderlinden, courtesy Waclaw Ustupski collection

Radom Rifle Production: G29/40 Rifle

(Fig. 248-1) G29/40 Rifle
Serial Number: ***1057***
Production Year: ***1939***
Caliber: ***7.92x57mm***
Notes: ***Captured Polish receivers were used at the Steyr factory (early war code 660) to produce German M98K Mauser short rifles, which were designated G29/40. The wz.29 marking on these receivers is usually altered by lining through the "wz" and adding a "g" but this was not done on this example. The front sight hood has been pushed back out of position.***
Photographs by A. Vanderlinden, courtesy Ron Pinkenburg collection

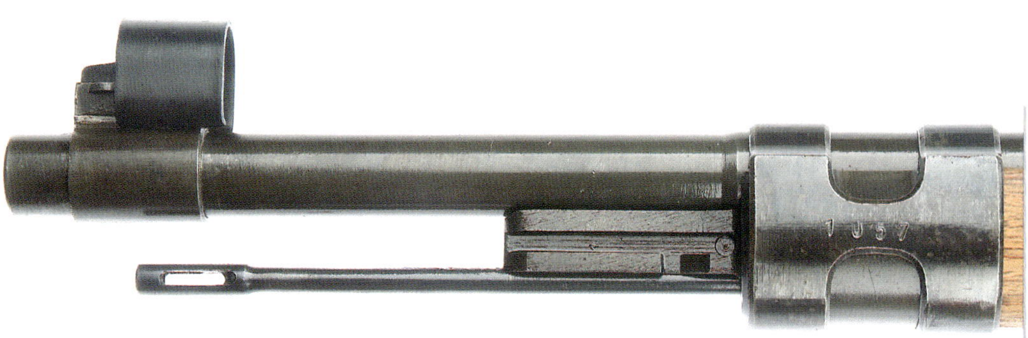

Radom Rifle Production: wz.1948 .22 Training Rifle

(Fig. 249-1) wz.1948 Training Rifle
Serial Number: ***ZM1108***
Production Year: ***1956***
Caliber: ***.22 LR***
Notes: ***Following the end of World War II, Poland was compelled to adopt Soviet small arms, including M38 and M44 Mosin-Nagant carbines. Accordingly, a .22 Long Rifle training rifle made to resemble the Mosin-Nagant carbines went into production. The action of this trainer is closely modeled after that of the wz.32 trainer illustrated above- the two trainers' bolts are nearly identical, but will not interchange. The number 11 in an oval atop the receiver ring of this rifle is the Communist era country code for Poland.***
Photographs by A. Vanderlinden, courtesy Waclaw Ustupski collection

(Fig. 250-1) German soldiers pose for this photograph in a dark humorous way, somewhere in Eastern Europe. One soldier is pointing a VIS. (Photograph courtesy of Anthony Vanderlinden)

(Fig. 251-1) A Russian soldier with a prewar Polish VIS holster. (Photograph courtesy of Waclaw Ustupski)

Bibliography

Books

Dr. Wlodzimier Antoniewicz, et al., Polska Jej Dzieje i Kultura od czasow najdawniejszych do chwili obecnej (3 vols.) , Warszawa, Traski, Evert,i,Michalski, 1927

Andrzej Albert (Wojciech Roszkowski), Najnowsza Historia Polski 1914-1993, Kielce, Exbud, Orion, Art-Kielce, 1994

Robert J. Berger, THE RADOM VIS PISTOL, R.J. Berger 1981

Eugene J. Bender, Military Holsters of World War II, Dallas, Taylor, 1984

Gerhard Bock, Moderne Faustfeuerwaffen und ihr Gebrauch, Neudamm, Neumann, 1941

Geoffrey Booyhroyd, THE HANDGUN, New York, Crown, 1970

Andrzej Cieplinski, Ryszard Wozniak, 9mm Pistolet samopowlarzalny VIS Wz. 1935, Warszawa, Bellona, 1996

Joachim Gortz and Geoffrey L. Sturgess, Pistole Parabellum (3 Vols.), Cobourg, Collector Grade, 2010

Zbigniew Gwozdz, Polskie konstrukcie broni strzeleckij, Warszawa, Sigma, 1993

Edward C. Ezell, Handguns of the World, Harrisburg, Stackpole, 1981

George B. Johnson and Hans Bert Lockhoven, International Armament (2 Volumes), Cologne, Interarms, 1965

Daniel W. Kent, German 7.9mm Military Ammunition, Ann Arbor, D.W. Kent, 1990

Stanislaw Komornicki, Zygmunt Bielecki, Wanda Bigoszewska, Adam Jonca, Wojsko Polskie 1939-1945, Barwa i Bron, Warszawa, Interpress, 1990

Klaus-Peter Konig and Martin Hugo, 9mm Parabellum, Atglen, Schiffer, 1992

Andrrej Konstankiewicz, Bron Strelecka i sprzet artyleryjski formacij polskich i Wojska Polskiego w latach 1914-1939, Lublin, Uniwersitet, 2003

Terrence W. Lapin, VIS: the Model 35 Radom Pistol, Arlington, Hyrax, 2004

Richard D. Law, Backbone of the Wehrmacht, Toronto, Collector Grade, 1991

Marian Maciejewski, Bron Strzelecka Wojsk Polskich w latach 1717-1945, Szezcin, Glob, 1991

Dieter H. Marschall, Handguns of the Armed Organizations of the Soviet Occupation Zone & German Democratic Rep., Los Alamos, UCross, 2003

J. Howard Matthews, Firearms Identification (3 Volumes), Springfield, Thomas, 1965

Ministerstwo Spraw Wojskowich, Pistolet VIS wz. 1935, Warszawa, 1937

NRA Firearms Assembly 4- The NRA Guide to Pistols and Revolvers, Fairfax, NRA, various ed.

W. H. B. Smith and Joseph E. Smith, The Book of Pistols and Revolvers, Harrisburg, Stackpole, 1968

Michael Steves and Bruce Karem, Kriegsmodell, Mentor, 3d. Party, 2010

Jan C. Still, Axis Pistols, Douglas, Still, 1986

Anthony Vanderlinden, FN Browning Pistols, Greensboro, Wet Dog Publications, 2009

Anthony Vanderlinden, Holsters and Shoulder Stocks of the World, Greensboro, Wet Dog Publications, 2005

Zdislaw Walczak, 9mm Pistolet wz 1935 VIS, Warszawa, M.O.N., 1983

Maj. Robert D. Whittington III, German Pistols and Holsters 1934-1945 (4 Volumes), Highland Park, Gun Room, 1976 (v.1); Hooks, Brownlee, 1989(v.2), 1990 (v.3), 1991 (v.4)

J. B. Wood, Troubleshooting Your Handgun, Chicago, Folett, 1978

Jacek Wolfram, Bron Strzelecka Armii Nemeckiej 1871-1945, Czesc I, Pistolety i Rewolwery, Gdansk, L&L, 2007

Adam Zamoyski, The Polish Way, London, Murray, 1987

Articles and Miscellaneous Materials

"Fabryka Broni w Radomiu w Latach 1922-1939", Biuletyn Kwartalny Radomskieg Towarzystwa Naukowego, tom xxxviii 2003 zeszyt 1-2

"The Polish Radom Pistol" by W. H. B. Smith, The American Rifleman, Nov, 1944, pp. 19&20

"The Radom…Toughest 9mm ever" by Jim Thompson, Gun Digest, 41st. ed., 1987, pp. 17-21

"VIS i MORS" by Piotr Wilniewczyc, Musealnictwo Wojskowe I (1959), pp.317-333

" Najcenniejsze Nabytki Muzeum Wojska Polskiego w Latach 1981-1986 " by Roman Matusewski and Mieczyslaw Sicinski, Musealnicstwo Wojskowe 4 (1989), pp 43-92

" Zagadka Kolby do Pistoletu VIS " by Leszek Komuda, Musealnictwo Wojskowe 6 (1992), pp. 122-139

" Poland's Unwilling Contributon to the Third Reich, Radom VIS " By Peter G. Kokalis, Shotgun News, July 10, 2006, pp. 12-16

" Pistolet bocznego zaplonu w Polsce " by Zbigniew Gwozdz, Strzal #11, 2009, pp. 56-59

" Archer " Radom Arms Factory LLC History at http://en.fb.pawelec.info/index.php?d=128

"Informacja Zaklady Metalowe LUCZNIK im. Gen. Waltera w Radomiu 1922-2000", mgr Malgorzata Comber, Radom, 2005

"Historia VIS-a", http://www.vis.zamosc.pl/stronavis/vis/history.htm

"Pistolet wz.35 Vis", http://www.wojsko18-39.internetdsl.pl/wrzesien/vis.html

"VIS wz. 1935 History Description Variants Users", Polish Firearms Page, http://hem2.passagen.se/dadkri/Vis35_nf.htm

2010 Radom Factory Brochure "VIS wz. 35 Radom Pistol Replica "

About the Photography

The photography illustrated throughout this book represents the pistols in their current condition. Images were prepared in order to best reproduce in printed form, pistol retouching was limited to obvious flaws like removal of some dust specs. No pistol photographs were retouched in order to enhance, improve, or alter the condition of the pistols. Some detail images were sometimes retouched to the extreme or altered in order to show/translate the features that are discussed, this was most notably done with small markings and holster markings/stamps.

The cover photograph was retouched to a higher degree, yet the illustrated pistols remain correct, representing their original condition.

All photographs were originally taken for black and white reproduction in the first edition. For the revised color edition; we have retouched the original photographs to reproduce color as accurately as possible. Despite our attempts to be true on color, we have not matched any photos to the original pistols as this was not feasible. Therefore photographs can vary in color from the actual pistol.

The cover photograph includes: (top to bottom)

Wartime German holster for the VIS (see page 82)
A 1918 dated Polonia Restituta (Poland Restored) medal
A set of three commemorative Polish stamps issued August 6, 1939;
  "25 Year Anniversary of March to Freedom"
A photograph of two Polish resistance members with VIS pistols
Late war VIS with brown Bakelite grips, second alphabet series N° G3973
Late war BNZ marked VIS pistol N° K2024 (see page 216)
Original VIS lanyard (see page 89)
A prewar Polish armband
A mint prewar Polish VIS pistol
Prewar VIS magazine (see page 30) and cleaning rod (see page 88)
Stock certificate from 1927 for one share in the Polish ammunition
  company "Pocisk" ("Bullet")
A (prewar) Polish cavalry hat (modern reproduction)
A (prewar) Polish officer's jacket (modern reproduction)
VIS prototype N° 0108 (see pages 114-116)
  with three rounds of wartime Polish 9x19mm ammunition
1934 Polish Ng30 Nagant revolver (see page 235)
Prewar VIS holster and officer's belt (see pages 79-82)

Cover page: Two Polish resistance fighters during the Warsaw Uprising, armed with a VIS and a P-38 pistol. (See pages 77-78) (Photograph courtesy of Waclaw Ustupski)

Back cover: Stamping of the Auto-Ersatzteilwerkstatte Znaim (Spare Parts Factory Znaim); the factory where the VIS was produced in 1945. (Photograph courtesy of Waclaw Ustupski)

About the Author

Bill York is a 74 year old retired U.S. Treasury Department lawyer who has since his early youth had an interest in guns, specifically in learning how unfamiliar ones work. He has sought out firearms with odd systems, learned how to disassemble them, and (mostly) reassembled them.

His heritage on his mother's side is Polish, and he has for some years been drawn to the construction and history of Poland's VIS (Radom) pistol, befriending some notable collectors along the way, and ultimately writing this book; a sort of historical photo album with commentary.

In 1990, when still a Federal Government employee, he had the opportunity of meeting Soviet weapon designer Mikhail T. Kalashnikov on his first trip to the United States. Since that time, he has visited Lieutenant General Kalashnikov in Russia on a number of occasions, and has served as a co-editor (with his friend Edward T. Costello) of an English language translation of Elena M. Kalashnikova's biographical work about her father, which was published by Thomas Nelson of Ironside Publications in 2011.

York collects more books than guns, and his library in the Polish as well as English languages has been of great service in the preparation of this book. York is a life member of the National Rifle Association, a charter life member of the Virginia Gun Collectors' Association, Inc., a member of the Ohio Gun Collectors' Association, and a member and past treasurer of Ye Connecticut Gun Guild.

Also from Wet Dog Publications

Altai is one of the Borzoi (Russian Wolfhound) Wet Dogs. He was rescued in 2019 in Barnaul, Siberia and traveled more than 7,200 miles to his new home in North Carolina. He is now one of the Wet Dog Publications company mascots and also represents the National Borzoi Rescue Foundation.

Please see our website at www.fnbrowning.com for our current selection of titles including details, availability, and pricing.

Our title selection:

Allied Rifle Contracts in America
Ars Mechanica, The Utimate FN Book
Browning Auto-5 Shotguns, The Belgian FN Production
FN-49, The Last Elegant Old-World Military Rifle
FN Browning Pistols, Side-Arms that shaped World History
FN Mauser Rifles, Arming Belgium and the World
Kongsberg Colten (Norwegian 1914)
Politiebewapening, Armament of the Dutch Police 1945-2005
VIS Radom, A Study and Photographic Album of Poland's Finest Pistol
Winchester Model 1895, Last of the Classic Lever Actions